James Henry Emerton

The Structure and Habits of Spiders

James Henry Emerton

The Structure and Habits of Spiders

ISBN/EAN: 9783743322424

Manufactured in Europe, USA, Canada, Australia, Japa

Cover: Foto ©berggeist007 / pixelio.de

Manufactured and distributed by brebook publishing software
(www.brebook.com)

James Henry Emerton

The Structure and Habits of Spiders

AMERICAN

NATURAL HISTORY SERIES.

THE
STRUCTURE AND HABITS
OF
SPIDERS.

BY

J. H. EMERTON.

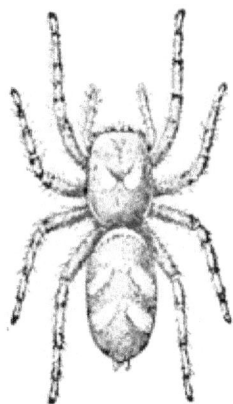

ILLUSTRATED.

SALEM:

S. E. CASSINO, PUBLISHER,

NATURALISTS' AGENCY,

1878.

Electrotyped
By C. J. Peters & Son,
Boston.

PREFACE.

———◆———

THE object of this book is to give a plain account of the best known habits of spiders, and as much of their anatomy and classification as is necessary to understand these habits. The portion on the spinning and flying habits is copied chiefly from Blackwall and Menge; that on the trap-door spiders from Moggridge; and the habits of Nephila and Hyptiotes, from Wilder. The observations of these authors have been repeated as far as possible, and some changes and additions made to their accounts of them. The numerous stories of deadly poison, supernatural wisdom, and enormous size and strength of spiders, have been omitted as doubtful. Several cuts from the papers of Professor Wilder have been repeated by favor of the author and publishers. Most of the figures are, however, new, and engraved by photography from my own drawings.

CONTENTS.

CHAPTER I.

CHAPTER IV.

LIST OF ILLUSTRATIONS.

THE STRUCTURE AND HABITS OF SPIDERS.

CHAPTER I.

ANATOMY AND CLASSIFICATION.

THE spiders form a small and distinct group of animals, related to the scorpions, the daddy-long-legs, and the mites, and less closely to the insects and crabs. They are distinguished by the more complete separation of the body into two parts; by their two-jointed mandibles, discharging a poisonous secretion at the tip; and by their spinning-organs, and habits of making cobwebs and silk cocoons for their eggs.

The common round-web spider, *Epeira vulgaris* of Hentz, will serve as well as any species to show the anatomy of spiders in general. Fig. 1 shows the under side of this

spider; Fig. 4, the upper side; and Fig. 5, an
imaginary section through the body, to show
the arrangement of the internal organs. To
begin with Fig. 1 : the body is seen to be di-
vided into two parts, connected only by the narrow joint, A, just behind the last pair of legs.
The front half of the body, called the thorax,
contains the stomach, the central part of the
nervous system, and the large muscles which
work the legs and jaws. The hinder half, the
abdomen, contains the intestine, the breathing-
organs, the principal circulating-vessels, the organs of reproduction, and the spinning-organs.
Connected with the thorax are six pairs of
limbs, four pairs of legs, B B B B, a pair of
palpi, C, and a pair of mandibles, D.

LEGS.

The legs are used chiefly for running, jumping, and climbing; but the front pair serve often
as feelers, being held up before the body while
the spider walks steadily enough on the other
six. One or both of the hinder legs are used to
guide the thread in spinning; the spider at the
same time walking or climbing about with the

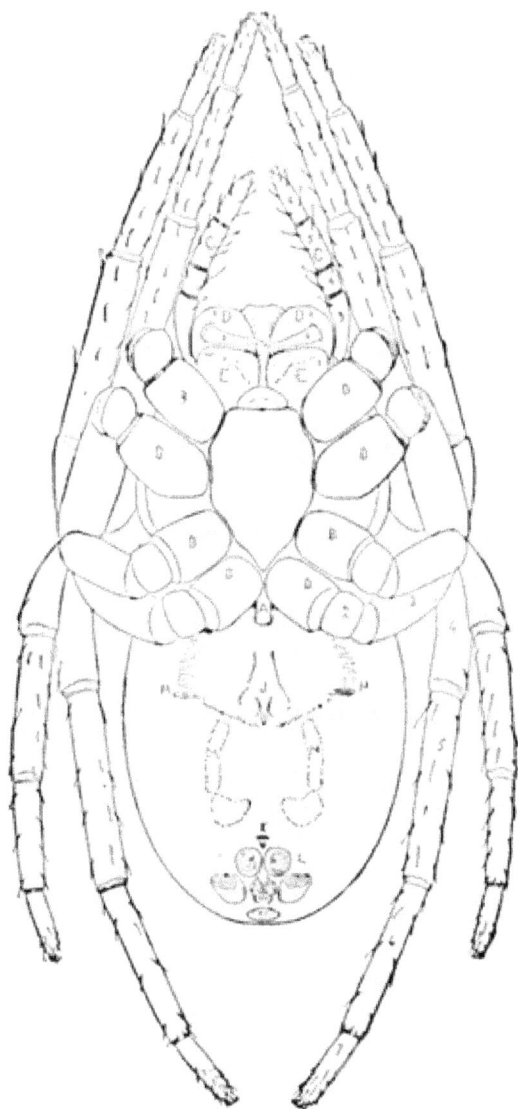

Fig. 1.

other six or seven. The legs are seven-jointed;
and on the terminal joint are three claws, Fig.
2, A, B, C, and various hair and spines. In
many spiders a brush of hairs takes the place
of the middle claw, as in the jumping spiders,
Fig. 3. Spiders with these brushes on their
feet can walk up a steep surface, or under a

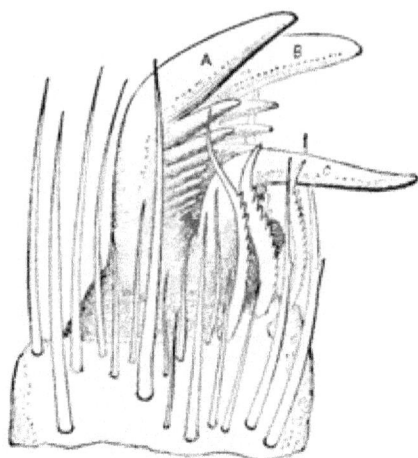

Fig. 2.

horizontal one, better than those who have
three claws. The legs of most spiders have
among the hairs movable spines, which, when
the spider is running about, extend outward
at a right angle with the leg, and, when it is
resting, are closed down against the skin.

PALPI.

In front of the legs are the palpi, Fig. 1, C, C,
— a smaller pair of limbs, with six joints and
only one claw or none. They are used as feelers,
and for handling food, and, in the males, carry

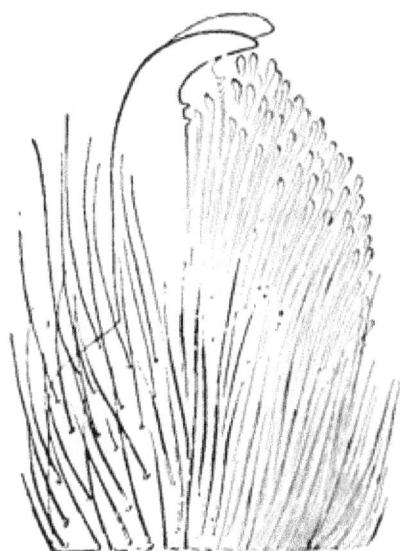

Fig. 3.

the curious palpal organs, which will be described farther on. The basal joints, Fig. 1,
E, of the palpi are flattened out, and serve as
chewing-organs, called "maxillæ."

Mr. Mason has lately described, in the Transactions of the Entomological Society of London, a large spider which has teeth on the inside of the palpi, which, when the spider is angry, are rubbed against teeth on the mandibles, producing a noise.

MANDIBLES.

The front pair of limbs, the mandibles, Fig. 1, D, are two-jointed. The basal joint is usually short and stout, and furnished on the inner side with teeth and hairs. The terminal joint is a small and sharp claw, which can be closed against the basal joint when not in use.

ABDOMEN.

On the under side of the abdomen, just behind the last pair of legs, are two hard, smooth patches, which cover the front pair of breathing-organs, the openings to which are two little slits at Fig. 1, H. Between these is the opening of the reproductive organs, and, in female spiders, the epigynum, Fig. 1, J, — an apparatus for holding the reproductive cells of the male.

At the end of the body are the spinnerets,

which will be described in another chapter.
There are three pairs of them; but many
spiders close them together when not in use,
so as to cover up the middle pair. The third
pair of spinnerets are often several-jointed,
and extend out behind the body like two tails.
In front of the spinnerets is a little opening, Fig. 1, K, which leads to air-tubes that give
off branches to different parts of the abdomen.
At M, Fig. 1, are usually two colored bands, or
rows of spots, marking the course of muscles
attached to the skin at various points along
these lines.

Fig. 4 is the back of the same spider. The
head is not separated from the rest of the body,
as in insects, but forms, with the thorax, one
piece. On the front of the head are eight eyes,
O, which are differently arranged in different
spiders. At the back part of the thorax is a
groove, P, under which is attached a muscle
for moving the sucking-stomach, Fig. 5, *d.*
From this point radiate shallow grooves, that
follow the divisions between the muscles of the
legs. On the abdomen are several pairs of
dark smooth spots, which mark the ends of

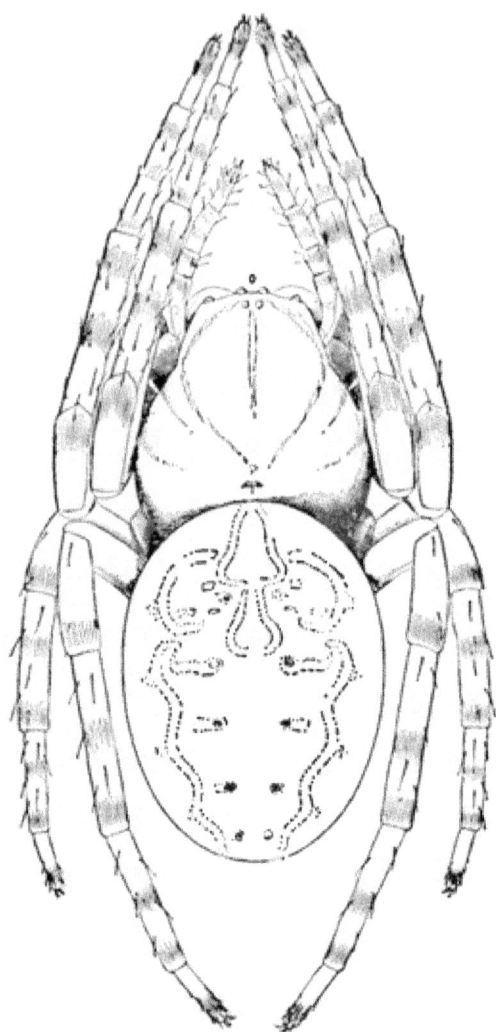

Fig. 4.

muscles extending downward through the abdomen. The markings of this spider are very complicated. The spot on the middle of the front of the abdomen is a very common one, and, in some spiders, extends the whole length of the body. The waved lines on each side are also common, and, in long-bodied spiders, often form two bright-colored stripes, or rows of spots, running nearly straight the whole length of the abdomen.

INTESTINE.

Fig. 5 is a section of the same spider. The mouth is at *a b*, just under and behind the mandibles, and between the maxillæ. It has an upper, *a*, and under lip, *b*, each lined with a horny plate, in the middle of which runs a groove. When the lips are closed, the two grooves form a tube, which leads to the œsophagus, *c*, and so into the stomach. At the end of the œsophagus is the sucking-stomach. This consists of a flattened tube, to the top of which is attached a muscle, *d*, connected with the groove in the back; and to the bottom, muscles, *f*, attached to a tough diaphragm spreading across

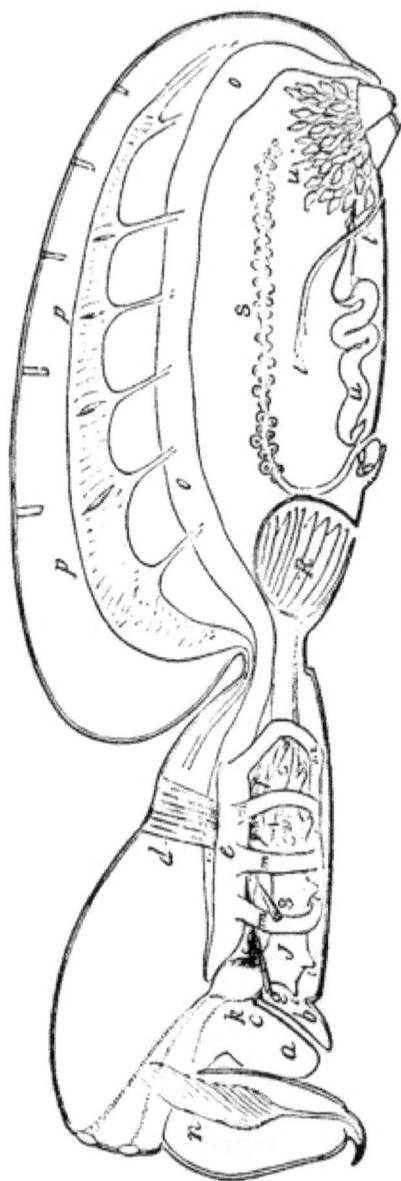

Fig. 5.

Section of a spider to show the arrangement of the internal organs: *a*, *b*, upper and under lips of the mouth; *c, c*, the œsophagus; *d, f*, upper and under muscles of the sucking-stomach; *e*, stomach; *g, g*, ligaments attached to diaphragm under the stomach; *J*, lower nervous ganglion; *k*, upper ganglion; *l, l*, nerves to the legs and palpi; *m*, branches of the stomach; *n*, poison-gland; *o*, intestine; *p*, heart; *R*, air-sac; *S*, ovary; *t*, air-tube; *u*, spinning-glands.

the thorax, and fastened between the legs on each side at *g g.* When these muscles contract, the top and bottom of the sucking-stomach are drawn apart, and whatever is in the œsophagus sucked in. By this pumping motion the spider is supposed to take liquid food from the mouth, and drive it backward into the abdomen. Just behind the sucking-stomach, the intestine gives off two branches, *c c*, which extend forward around the stomach muscle, and meet over the mouth. Each of these branches gives off on the outer side four smaller branches, *m m m m*, which extend downward, — one in front of each leg, — and unite on the under side of the thorax.

The intestine, *o*, continues backward through the abdomen to the anus, in the little knob behind the spinnerete. The brown mass which surrounds the intestine, and fills the abdomen above it, is supposed to be a secreting-organ discharging into the intestine at several points.

HEART.

Over the intestine, and parallel with it, is the heart, *p*, a muscular tube, with openings along the sides to receive the blood, and branches through which it flows to different parts of the body. The greater part of the blood enters at the front of the heart, and passes backward into the abdomen, or forward into the thorax.

BREATHING-ORGANS.

In the front of the abdomen are the principal breathing-organs, — a pair of sacs, R, containing a number of thin plates, through which the blood passes on its way to the heart. Besides these, there is a pair of branching air-tubes, *t*, opening near the spinnerets.

NERVOUS SYSTEM.

The nervous system has a large ganglion, J, in the thorax, from which branches, *i*, pass to the limbs and abdomen. At the front end two branches extend upward, each side of the œsophagus, to two smaller ganglia, *k*, from which pass nerves to the mandibles and eyes.

The reproductive organs, S, lie along the under side of the abdomen, and open between the two air-sacs.

The spinning-glands, *u*, lie above the spinnerets, and along the under side of the abdomen. They will be more fully described in the chapter on spinning.

POISON-GLANDS.

The poison-glands, *n*, are partly in the basal joints of the mandibles, and partly in the head, and discharge by a tube which opens at the point of the claw of the mandible, Fig. 15, *a*.

CLASSIFICATION.

There is not room in this book to explain the classification of spiders into genera and species; but a description of the following well-marked groups, which contain nine-tenths of all spiders, will give a general idea of the differences among them, and help to understand what follows.

MYGALIDÆ.

This family includes the largest known spiders. The body is usually very hairy and dark-colored. Most species have only four spinnerets; and one pair of these are long, and are turned up behind the abdomen. They have

four air-sacs under the front of the abdomen,
instead of two, as other spiders.   Their man-

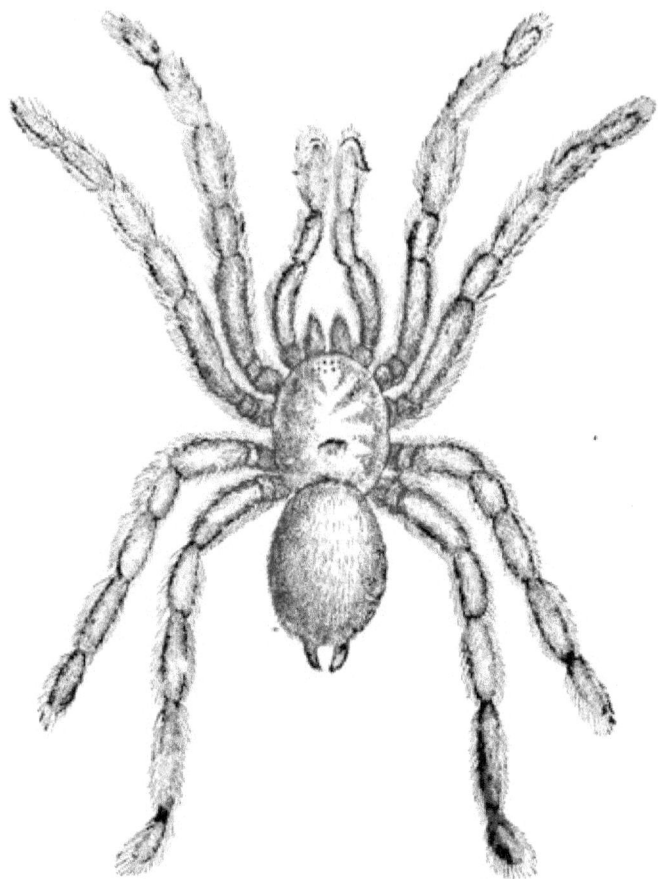

Fig. 6.

dibles are very large, and work up and down,
instead of sidewise.   The eyes are collected

together on the front of the head. They live only in warm countries. Specimens from South America are exhibited in every natural history museum. Fig. 6 represents *Mygale Hentzii*, a species living in Arizona and Texas.

DYSDERIDÆ.

A small family of spiders with only six eyes. They have also four breathing-holes in the front of the abdomen ; but one pair leads to branched tubes instead of sacs. They are usually found under stones, with their legs drawn up close to their bodies, but can move very quickly when so inclined. Very few species are known, and none are common, in North America. Fig. 7 is *Dysdera interrita* enlarged. Below are the eyes as seen from in front.

DRASSIDÆ.

A large family of spiders, varying greatly in shape, color, and habits. Most of them are dull colored, and live under stones, or in silk tubes on plants, and make no webs for catching insects. Their eyes are small, and arranged in two rows on the front of the head. Their feet have two claws and a bunch of flat hairs. The spinnerets are usually long enough to extend a little behind the abdomen. Fig. 8 is a *Drassus*, and the eyes as seen from in front.

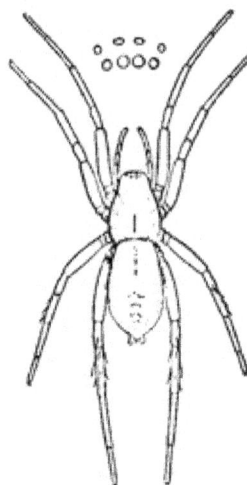

Fig. 7.   Fig. 8.

AGALENIDÆ.

Long-legged, brown spiders, with two spinnerets longer than the others, and extending out behind the body. They make flat webs, with a funnel-shaped tube at one side, Fig. 24, in which the spider waits. Fig. 9 is *Agalena nævia*, the common grass spider.

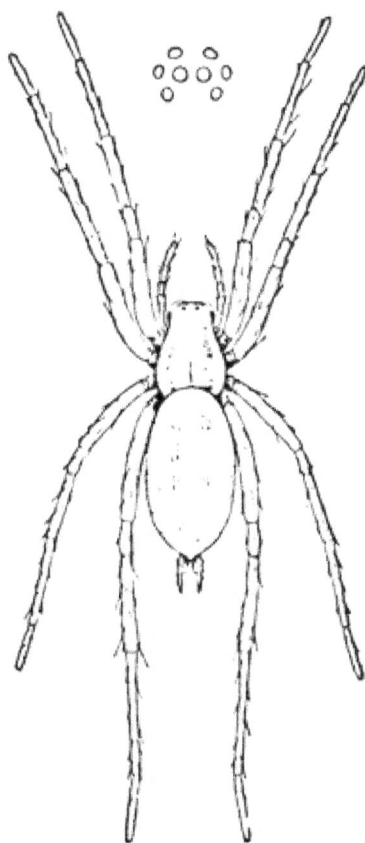

Fig. 9.

CINIFLONIDÆ.

A family resembling the last, but having peculiar spinning-organs, which will be described farther on, Fig. 35.

LYCOSIDÆ.

The running spiders. Very active spiders,

living in open places, and catching their prey without webs. Their legs are long, the hind pair being the longest. The head is high, and the arrangement of the eyes peculiar to the family, Fig. 10.

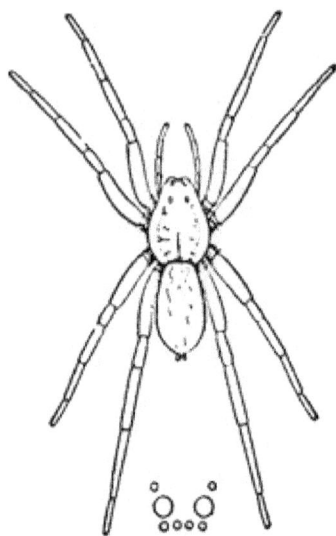

Fig. 10.

ATTIDÆ.

The jumping spiders. The body is usually short, and the head square. The arrangement of the eyes is characteristic, Fig. 11, *a*. The two large ones in the middle give these spiders a more animated look than those of other families. The legs are short, and the front pair often stouter than the others. They can walk easily backwards or sidewise, and jump a long distance. Fig. 11 is the common gray jumping spider enlarged.

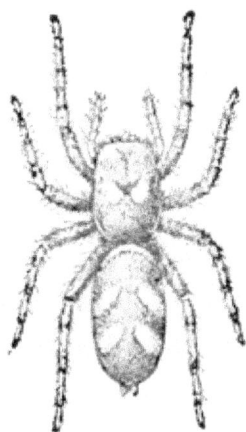

Fig. 11 *a.*

Fig. 11.

THOMISID.E.

The crab spiders. The body is usually flat, and wide behind. The front two pair of legs are longer than the others, and so bent that the spider can use them when in a narrow crack. Some of them, like crabs, walk better sidewise than forwards. Like the running and jumping spiders, they make no webs for catching food.

The eyes are small, and simply arranged in two rows, Fig. 12.

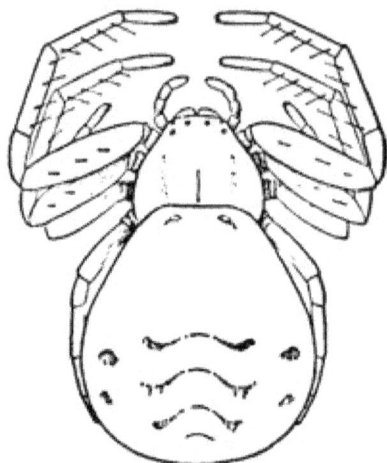

Fig. 12.

THERIDIID.E.

The largest family of spiders.   Most of them are small, with large rounded abdomens and slender legs.   They live usually upside down, holding by their feet under their webs.   They make large cobwebs of different shapes for different species, and depend for food on what is caught in them.   To this group belongs the genus *Erigone*, containing a great number of the smallest spiders known. Fig. 13 is the common house *Theridion* enlarged.

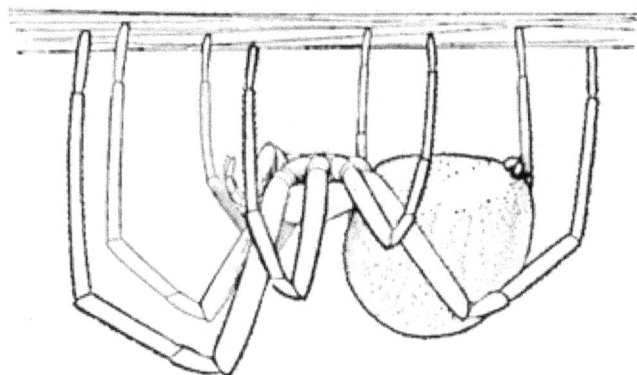

Fig. 13.

EPEIRIDÆ.

The round-web spiders. Large spiders, with flat heads, and eyes wide apart, and short, round abdomens, Figs. 1, 4. They make webs formed of radiating lines crossed by other adhesive ones in a spiral or concentric loops, Fig. 28. They hang in the web, head downward, or live upside down in a hole near by.

CHAPTER II.

THE spiders are best known and hated as animals that bite. The biting apparatus is shown in Fig. 14, which represents the head and mandibles of *Epeira vulgaris*, seen from in front. When not in use, the claw is closed up against the mandible between the rows of teeth ; but, when the jaws are opened to bite, the claws are turned outward, so that their points can be stuck into any thing between the jaws. Fig. 15 is the claw still more enlarged, showing a little hole near the point at *a*, out of which is discharged the secretion of a gland in the head, Fig. 5, *n*. The ordinary use of the mandibles is for killing and crushing insects, so that the soft parts can be eaten by the spider ;

32

and in this they are aided by the maxillæ, Fig.
1, E. They will sometimes chew an insect for
hours, until it becomes a round lump of skin,

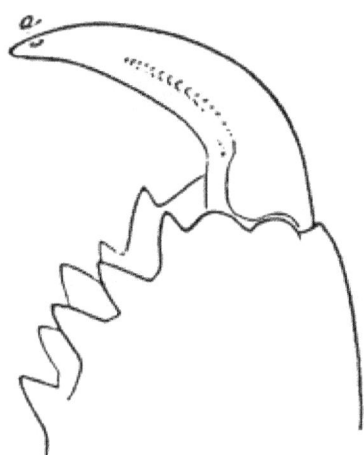

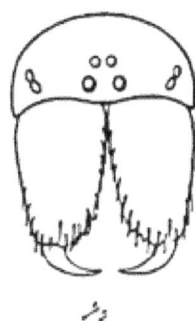

Fig. 14.   Fig. 15.

with all the blood sucked out of it ; this is then
thrown away, the spider swallowing only such
bits as may happen to be sucked in with the
liquid portion.

If let alone, no spiders bite any thing except
insects useful for food ; but, when attacked and
cornered, all species open their jaws, and bite
if they can ; their ability to do so depending
on their size, and the strength of their jaws.
Notwithstanding the number of stings and
pimples that are laid to spiders, undoubted

cases of their biting the human skin are very rare; and the stories of death, insanity, and lameness from spider-bites, are probably all untrue.

Many experiments have been tried to test the effect of the bites of spiders on animals. Doleschall shut up small birds with *Mygale Javanica* and *Mygale Sumatrensis*, both large and strong spiders; and the birds died in a few seconds after being bitten. One of the spiders was left for ten days without food, and then made to bite another bird, which was injured, but in six hours recovered. The same author was bitten in the finger by a jumping spider. The pain was severe for a few minutes, and was followed by lameness of the finger, and gradually of the hand and arm, which soon went away entirely.

Bertkau allowed spiders to bite his hand. On the ends of the fingers the skin was too thick; but between the fingers they easily pricked it. The bite swelled and smarted for a quarter of an hour, and then itched for some time, and for a day after itched whenever rubbed, as mosquito-bites will. He also experimented on flies, which died in a few minutes after being bitten.

Mr. Blackwall, to test the poison of spiders, made several large ones bite his hand and arm, and at the same time pricked himself with a needle. Although the spiders bit deep enough to draw blood, the effect of their bite was exactly like that of the prick of the needle. No inflammation or pain followed, and both healed immediately.

Several spiders were placed together, and made to bite one another. The bitten ones lived always some hours, and died from loss of blood ; and one spider, that had been bitten in the abdomen so that some of the liver escaped and dried on the outside, lived over a year, apparently in good health.

A large spider was made to bite a wasp near the base of the right front-wing, so as to disable it ; but it lived thirteen hours.

A bee was bitten by a large spider, but lived three days.

A grasshopper was bitten, and held in the jaws of a spider for several seconds ; but it lived in apparent health for two days.

Insects of the same kinds were wounded in the same places with needles, and died in about the same time as when bitten.

From these experiments Mr. Blackwall was led to believe that the secretion from the spider's jaws is not poisonous, but that insects die, when bitten, from loss of blood and mechanical injury.

Mr. Moggridge, who studied the habits of trap-door spiders for several years, was more than once bitten by them, but never had any pain or inflammation from the bites.

The bites of *Latrodectus guttatus* of the south of Europe, and an allied species in California, are much dreaded, but probably as much on account of the size and conspicuous colors of the spider as any thing else.

The Tarantula, also a south European spider, has been supposed to cause epilepsy by its bites, which could only be relieved by music of particular kinds. These stories appear, however, to be all nonsense: at any rate, the Tarantula bites produce no such effect nowadays. These spiders live in holes in sand, out of which they rush after passing insects, and may be caught by a straw moved carefully over the holes like an insect. They are no more savage in their habits than other spiders; and Dufour kept one that soon learned to take flies from his fingers without biting him.

Spiders of very different species soon learn to take food from the hand or a pair of forceps, or water from a brush, and will come to the mouth of their bottle, and reach after it on tip-toe.

Many stories are told of spiders coming out of their holes to listen to music, and of their being taught to come out and take food at the sound of an instrument.

CHAPTER III.

THAT which, more than any thing else, distinguishes spiders from other animals is the habit of spinning webs. Some of the mites spin irregular threads on plants, or cocoons for their eggs; and many insects spin cocoons in which to pass through the change from larva to adult. In the spiders the spinning-organs are much more complicated, and used for a greater variety of purposes, — for making egg-cocoons, silk linings to their nests, and nets for catching insects. The spider's thread differs from that of insects, in being made up of a great number of finer threads laid together while soft enough to unite into one.

38

SPINNERETS.

The external spinning-organs are little two-jointed tubes on the ends of the spinnerets, Fig. 1, L. Fig. 16 is the spinnerets of the same spider, still more enlarged to show the arrangement of the tubes. There is a large number of little

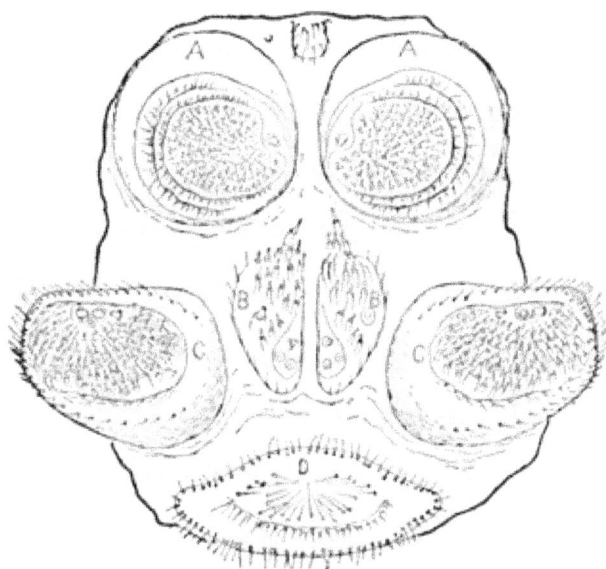

Fig. 16.

tubes on each spinneret, and in certain places a few larger ones. Fig. 17 is a single tube, showing the ducta which leads the viscid liquid to form the thread from a gland in the spider's abdomen. Each tube is the outlet of a separate gland. Fig. 18, *a*, shows four small tubes from a spinneret of *Epeira*, each with a small gland attached; and Fig. 18, *b*, a large tube, with one of the large glands which extends forward the whole length of the abdomen, Fig. 5, *u*.

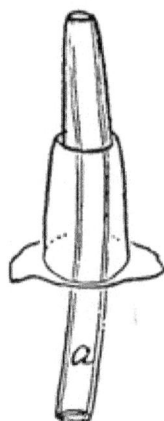

Fig. 17.   Fig. 18.

The shape of the spinnerets, and size and arrangement of the tubes, vary in different species. Fig. 19 is a spinneret of *Prosthesima*, where there are a few large tubes in place of many small ones. In *Agalena* the two hinder spinnerets are long, and have spinning-tubes along the under side of the last joint, Fig. 20.

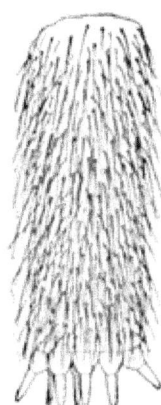

Fig. 19.

When the spider begins a thread, it presses the spinnerets against some object, and forces out enough of the secretion from each tube to adhere to it. Then it moves the spinnerets away; and the viscid liquid is drawn out, and hardens at once into threads, — one from each tube. If the spinnerets are kept apart, a band of threads is formed; but, if they are closed together, the fine threads unite into one or more larger ones. If a spider is allowed to attach its thread to glass, the end can be seen spread out over a surface as large as the ends of the spinnerets, covered with very fine threads pointing toward the middle, where they unite, Fig. 21.

The spinning is commonly helped by the hinder feet, which

Fig. 20.

guide the thread, and keep it clear of sur-
rounding objects, and even pull it from the
spinnerets. This is well seen when an insect
has been caught in a web, and the spider is
trying to tie it up with threads. She goes as
near as she safely can, and draws out a band
of fine threads, which she reaches out toward

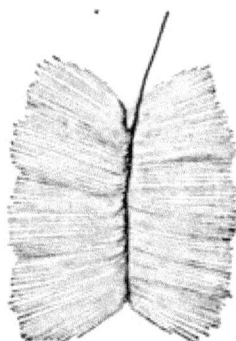

Fig. 21.

the insect with one of her hind-feet; so that
it may strike the threads as it kicks, and
become entangled with them. As soon as the
insect is tied tightly enough to be handled, the
spider holds and turns it over and over with her
third pair of feet, while, with the fourth pair,
she draws out, hand over hand, the band of
fine threads which adhere to the insect as it
turns, and soon cover it entirely.

It is a common habit with spiders to draw out a thread behind as they walk along; and in this way they make the great quantities of threads that sometimes cover a field of grass, or the side of a house. We often see the points of all the pickets of a fence connected by threads spun in this way by spiders running down one picket, and up the next, for no apparent purpose.

Spiders often descend by letting out the thread to which they hang; and are able to control their speed, and to stop the flow of thread, at will. They sometimes hang down by a thread, and allow themselves to be swung by the wind to a considerable distance, letting out the thread when they feel they are going in the right direction.

Spiders in confinement begin at once to spin, and never seem comfortable till they can go all over their box without stepping off their web. The running spiders, that .make no other webs, when about to lay their eggs, find or dig out holes in sheltered places, and line them with silk. Species that live under stones or on plants all line their customary hiding-places with web, to which they hold when at rest. Several of the large running spiders dig holes in sand, and line them with web, so that the sand cannot fall in; and build around the mouth a ring of sticks and straws held together by threads.

TRAP-DOOR NESTS.

The building of tubular nests is carried to the greatest perfection by certain genera of the *Mygalidæ.* (See page 13.)

Atypus, the most northern genus of this family, makes a strong silken tube, part of which forms the lining of a hole in the ground, and part lies above the surface, among stones and plants, Fig. 22, A. The mouth of the tube is almost always closed, at least when the spider is full grown.

Another genus, which lives in warm countries, makes tubes lined with silk, and closed at the top by a trap-door. A common species, *Cteniza Californica*, lives in the southern part of California, and is often brought east by travellers. It digs its hole in a fine soil, that becomes, when dry, nearly as hard as a brick; but the spider probably works when the ground is wet.

The holes are sometimes nearly an inch in diameter, and vary in depth from two or three inches to a foot. The mouth is a little enlarged, and closed by a thick cover that fits tightly into it, like a cork into a bottle. The cover is made of dirt fastened together with threads, and is lined, like the tube, with silk, and fastened by a thick hinge of silk at one side, Fig. 22, B. When the cover is closed, it looks exactly like the ground around it. The spider holds on the inside of the door with the mandibles and the two front pairs of feet; while the third and fourth pairs of legs are pressed out against the walls of the tube, and hold the spider down so firmly, that it is impossible to raise the cover without tearing it.

Among the trap-door spiders of Southern Europe, about which Mr. J. T. Moggridge has written a very interesting book, are species which make different kinds of nests. The cover, instead of being thick, and wedged into the top of the tube like a stopper, is thin, and rests on the top of the hole, Fig. 22, C, and is covered with leaves, moss, or whatever happens to be lying about; so that it is not easily seen.

Two or three inches down the tube is another door, Fig. 22, E, hanging to one side of the tube when not in use; but, when one tries to dig the spider out from above, she pushes up the lower door, so that it looks as if it were the bottom of an empty tube.

Another species digs a branch obliquely upward from the middle of the tube, closed at the junction by a hanging-door, which, when pushed upward, can also be used to close the main tube, Fig. 22, F. What use the spider makes of such a complicated nest, nobody knows from observation; but Mr. Moggridge supposes that when an enemy, a parasitic fly, for instance, comes into the mouth of the tube, the spider stops up the passage by pressing up against the lower door; but, if this is not enough, it dodges into the branch, draws the door to behind it, and leaves the intruder to amuse himself in the empty tube. The branch is sometimes carried up to the surface, where it is closed only by a few threads; so that, in case of siege, the spider could escape, and leave the whole nest to the enemy.

In these nests the spiders live most of the

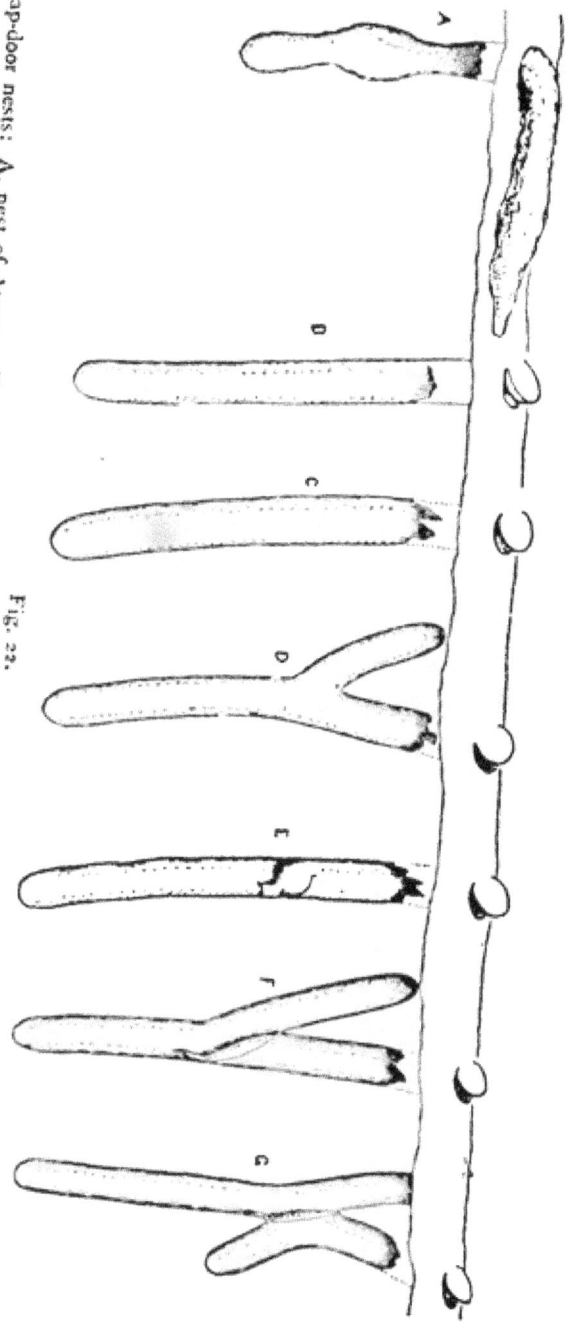

Fig. 22.

Trap-door nests: A, nest of Atypus; B, nest with thick door; C, nest with thin door; D, branched nest; E, nest with two doors; F, branched nest with two doors; G, nest with two branches.

time, coming out at night, and some species in the daytime, to catch insects, which they carry into the tube, and eat. The eggs are laid in the tube ; and the young are hatched, and live there till able to go alone, when they go out, and dig little holes of their own. As the spider gets larger, the hole is made wider, and the cover enlarged by adding a layer of earth and silk ; so that an old cover is made up of a number of layers, one over the other, over the original little cover.

Moggridge once took a *Cteniza Californica* out of her nest, and put her on a pot of earth, and the next morning had the good luck to see her at work digging. She loosened the earth with her mandibles, and took it in little lumps with the mandibles and maxillæ, and carried it away piece by piece. It took her an hour to dig a hollow as large as half a walnut. He saw the making of the door twice by other species. Once he dug a hole for a spider in some earth, and the next day found her in it, and the top covered by a little web, on which were scattered bits of earth and leaves, which had evidently been put there by the spider. The second

night, enough dirt and silk were added to make
the door of the usual thickness ; but the spider
never finished it so that it would open properly
on its hinge. Another time Moggridge saw at
the mouth of a very small hole a spider at work
making a door. She spun a few threads across
the hole, then gathered up with her front-legs
and palpi an armful of dirt, and laid it on top of
them. She then got under the pile, into the
tube ; but the motions of the dirt showed that
she was still at work on it, and next morning
the under side had been thickly covered with
web, and the whole separated from the mouth
of the tube, except at one side, where the usual
hinge was left. The new door was at first soft,
but in two or three days hardened, and appeared
exactly like an old door.

These spiders are accustomed to put on the
door moss like that which grows around it,
and so conceal the door from sight ; but when
Mr. Moggridge took away the moss, and dug up
the ground around a hole, and then destroyed
the cover, the spider made a new one, and
brought moss from a distance to put on it,
thereby making it the most conspicuous thing
in the neighborhood.

Mr. S. S. Saunders tried to see trap-door spiders make their nests. When the earth was dry, they would do nothing; but, after watering it, they several times dug new holes, but always in the night.

The food of the European trap-door spiders consists largely of ants and other wingless insects, and they have been known to eat earth-worms and caterpillars. Mr. Moggridge has often seen them, even in the daytime, open their doors a little, and snatch at passing in-sects, sometimes taking hold of one too large to draw into the tube. One time he and some friends marked some holes, and went and watched them in the night. The doors were slightly open, and some of the spiders' legs thrust out over the rim of the hole. He held a beetle near one of the spiders; and she reached the front part of her body out of the tube, push-ing the door wide open, seized the beetle, and backed quickly into the tube again, the door closing by its own weight. Shortly after, she opened it again, and put the beetle out alive and unhurt, probably because it was too hard to eat. He next drove a sow-bug near another hole;

and the spider came out and snatched it in the same way, and kept it. None of the spiders came entirely out of their holes, and they were only a little more active than in the daytime.

Erber, in the Island of Tinos, noticed a place where several trap-door nests were near each other, and spent a moonlight night watching them. Soon after nine o'clock some of the spiders came out, fastened back their doors, and each spun a web, about six inches long and an inch high, among the grass near her hole, and went back into the tube. In course of time beetles were caught in the webs, and eaten by the spiders, and the hard parts carried several feet from the nest. The next morning the webs had been cleared away, and the doors of the tubes closed, leaving no traces of the night's work.

SILK TUBES AND NESTS.

Several species of *Theridion* and *Epeira* make tents near their webs, under which they hang when at rest, and in which some species make their cocoons, and lay their eggs. The tents are usually covered outside with leaves drawn together, with sticks or bud-scales collected near by, or with earth and stones brought up from the ground below.

Some spiders living on plants make flat tubes, in which they wait for insects, and also hide while moulting, or laying eggs. Others make, especially about the breeding-time, bags

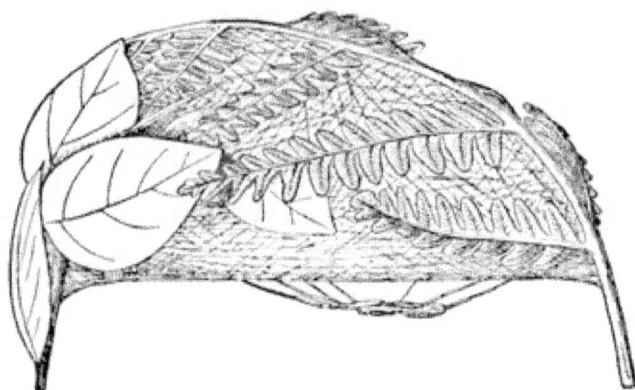

Fig. 23.

of silk on plants, or under stones, in which the egg-cocoons are finally spun.

Dolomedes makes among grass and shrubs, in meadows, a great nest, four or five inches in diameter, Fig. 23, in which is the egg-cocoon. The young hatch and ramble about in this nest for some time. The spider remains near, usually holding on under the nest.

THE WATER-SPIDER.

There is one spider that makes a bag of silk, something like those just mentioned, on water-plants, and lives in it under water, as in a diving-bell; the opening being below, so that the air cannot escape. Mr. Bell, in " The Journal of the Linnæan Society," vol. i., 1857, describes the filling of these nests with air by the spider. After the nest had been made as large as half an acorn, she went to the surface, and returned, fourteen times successively, and each time brought down a bubble of air, which she let escape into the nest. The bubble was held by the spinnerets and two hind-feet, which were crossed over them; and the method of catching it was the following : The spider climbed up on threads or plants nearly to the surface, and put the end of the abdomen out of water for an instant, and then jerked it under, at the same time crossing the hind-legs quickly over it. She then walked down the plants to her nest, opened her hind-feet, and let the bubble go.

The water-spiders run about on water-plants, and catch the insects which live among them.

They lay their eggs in the nest ; and the young come out, and spin little nests of their own, as soon as they are big enough. Their hairs keep the skin from becoming wet as they go through the water ; and in the nest they are as dry as if it were under a stone, or in a hole on land.

COBWEBS.

The simple nests and tubes that have been described are made by spiders, most of which spin no other webs. The larger and better known cobwebs for catching insects are made by comparatively few species. On damp mornings in summer the grass-fields are seen to be half covered with flat webs, from an inch or two to a foot in diameter, which are considered by the weatherwise as signs of a fair day. These webs remain on the grass all the time, but only become visible from a distance when the dew settles on them. Fig. 24 is a diagram of one of these nests, supposed, for convenience, to be spun between pegs instead of grass. The flat part consists of strong threads from peg to peg, crossed by finer ones, which the spider spins with the long hind-spinnerets, Fig. 20, swinging them from side to side, and laying down a

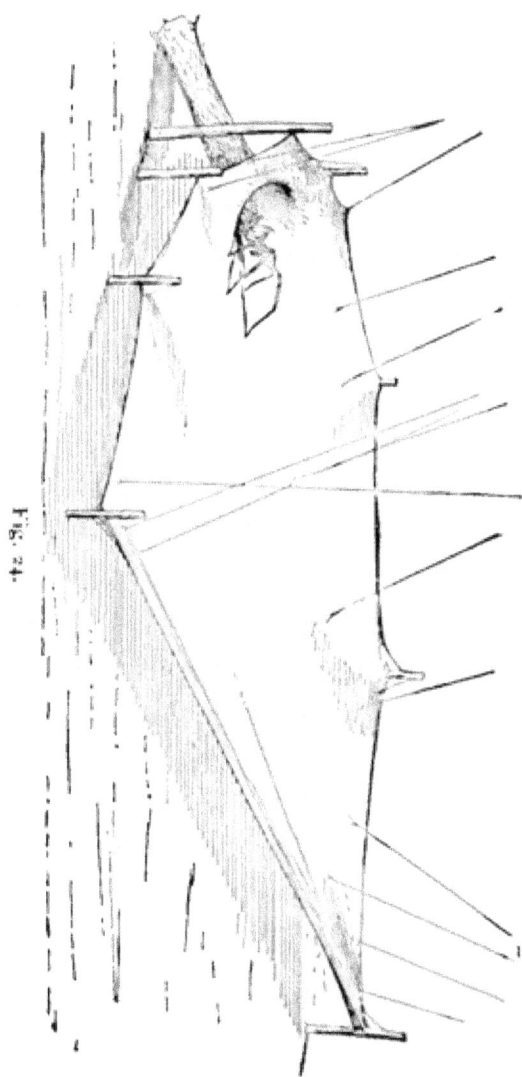

Fig. 24.

band of threads at each stroke. The web is so
close and tight, that one can hear the footsteps

of the spider as she runs about on it. At one side of the web is a tube leading down among the grass-stems. At the top the spider usually stands, just out of sight, and waits for something to light on the web, when she runs out, and snatches it, and carries it into the tube to eat. If any thing too large walks through the web, she turns around, and retreats out of the lower end of the tube, and can seldom be found afterward. In favorable places these webs remain through the whole season, and are enlarged, as the spider grows, by additions on the outer edges, and are supported by threads running up into the neighboring plants. Similar webs are made by several house-spiders, and are enlarged, if let alone, till they are a foot or two feet wide, and remain till they collect dirt enough to tear them down by its weight.

Nearly all spiders that make cobwebs live under them, back downward ; and many are so formed, that they can hardly walk right side up. The spiders of the genus *Linyphia* make a flat or curved sheet of web, supported by threads above and below ; the spider standing, usually, underneath in some corner, out of

sight.   *Linyphia Marmorata* makes a dome-
shaped web, Fig. 25, supported by threads that
extend up into the bushes two or three feet.

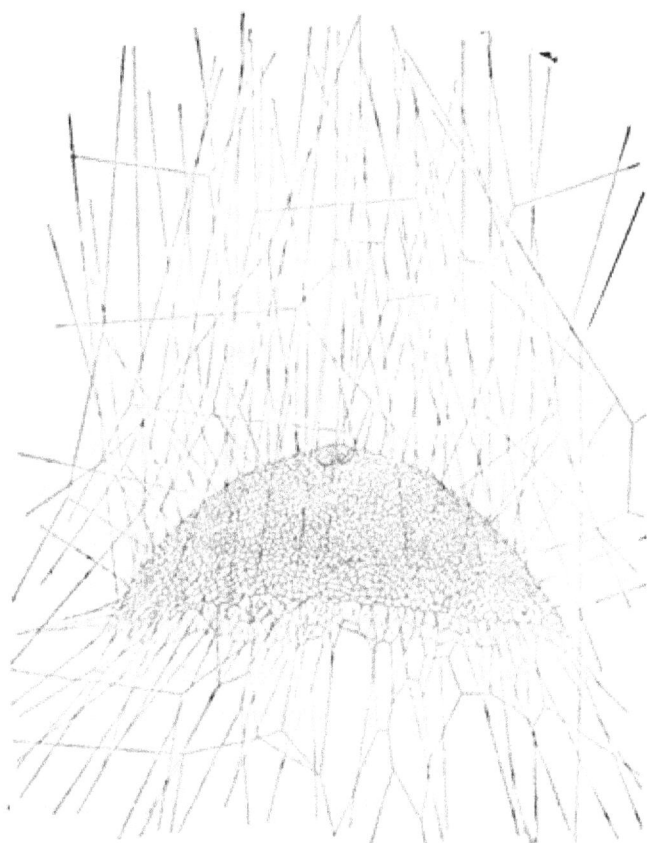

Fig. 25.

The spider stands under the middle of the
dome, where it draws in a small circle of web
with its feet.   The upper threads of the web in-

terfere with the wings of small insects flying
between them, and they fall down to the dome
below, where they are seized, and pulled
through the nearest hole. *Linyphia communis*
makes a double web, Fig. 26. The spider
stands under the upper sheet, which curves a
little downward. What the use of the lower
web is, is not easily seen. Either of these
spiders, when frightened, leaps out of the web
to the ground; but *Linyphia communis* must
go to the edge before she can clear herself,
and so is easily caught in her own web.

A little spider, *Argyrodes*, belonging to the
same family, lives among the upper threads of
webs of this kind, without being troubled by
the owner. It resembles in size and color the
scales of pine-buds that often fall in the web,
and may easily be mistaken for them. It
probably spins a few threads of its own among
the borrowed ones, and does, at times, make a
separate web of its own.

The webs of *Theridion* usually have at some
part a tent, or at least a thicker portion, under
which the spider stands; and from this run
irregularly simple threads, crossing each other

in all directions, and held in place by threads above and below. Such irregular webs are

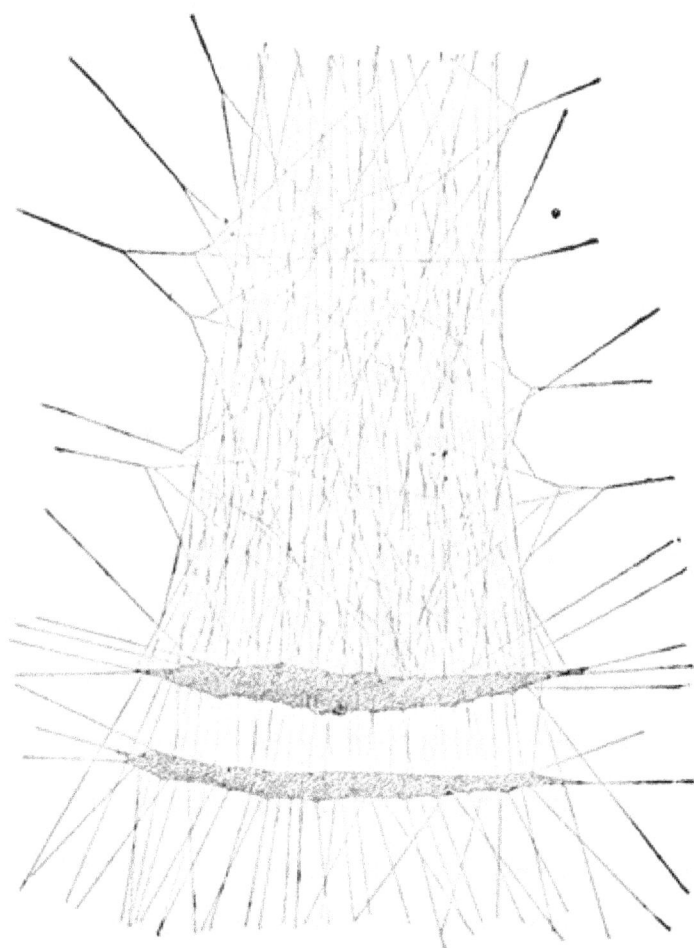

Fig. 26.

made often in houses by *Theridion vulgare,* Hentz, in corners of rooms, under furniture, and in cellar-stairways. The same spider spins occasionally out of doors on fences, but never on plants. When it has caught an insect, and tied it up, it hoists it up into the web, sometimes a considerable distance.

They do this by fastening to it threads from above, which, as they dry, contract, and pull it up a little. They keep on bringing down more and more threads, until the insect is at last hoisted to the top of the web, where they can suck it without exposing themselves.

Pholcus, the long-legged cellar-spider, makes an irregular web of this kind, and has a curious habit when alarmed. It hangs down by its long legs, Fig. 27, and swings its body around in a circle, so fast that it can hardly be seen. Fig. 27, *a*, represents the spider as seen from below; and the dotted circle shows the path in which it revolves.

ROUND WEBS.

These well-known cobwebs are made by the family *Epeiridæ*, Figs. 1, 4; and the process of making them by the common spider, from which these figures are drawn, can be easily

observed in any garden. They generally choose
for their web a window-frame or fence, or some
such open wooden structure, where there is a
hole or crack in which they can hide in the
daytime.

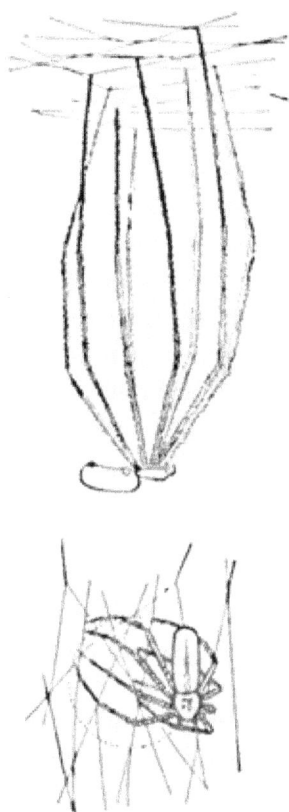

Fig. 27.

The spider begins by spinning a line across
where the web is to be, and attaches another
to it near the middle. She carries the last line

along, holding it off with one of the hind-feet,
and makes it fast an inch or two from one end
of the first; then she goes back to the centre,

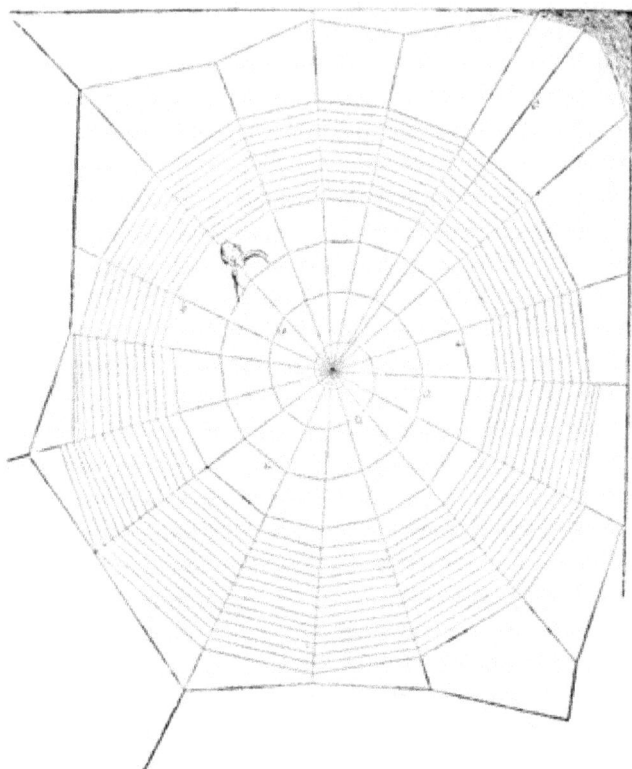

Fig. 28.

attaches another line, which she carries off in
another direction, and fastens; and so on, until
all the rays of the web, Fig. 28, are finished.

She stops occasionally at the centre, turns around, and pulls at the threads one after another, and spins here and there short cross-lines to hold them more firmly. She seems, by thus feeling the rays, to decide where to put in the next one, and does it always in such a way as to keep tight what has been done before. When the rays are finished to her satisfaction, the spider begins at the centre to spin a spiral line across them, Fig. 28, *a, a, a;* the turns of the spiral being as far apart as the spider can conveniently reach. She climbs across from one ray to the next, holding her thread carefully off with one of the hind-feet, till she gets to the right point, and then turns up her abdomen, and touches the ray with her spinnerets, thus fastening the cross-thread to it. The figure shows her in this position. When this spiral has been carried to the outside of the web, the spider begins there another and closer one, Fig. 28, of thread of a different kind. While the first thread was smooth, the latter is covered with a sticky liquid, which soon collects on it in drops, and makes it adhere to any thing that touches it.

After going round a few times, this spiral crosses the one that was spun first, or would, if the spider allowed it to; but, as she comes to the old spiral, she bites it away, leaving only little rags, Fig. 22, *b*, attached to the rays, which may be seen in the finished web. By beginning thus at the outside, the spider is able to cover the whole web with adhesive threads, and, without stepping on it, take her usual place in the centre. She usually is careful enough to spin beforehand a thread from the .centre to her nest, and sometimes stays there, with one foot on the thread, so as to feel if any thing is caught in the web. When she feels a shake, she runs down to the centre, feels the rays to see where the insect is, and runs out, and seizes it, or ties it up as described on page 43. We have described the web as consisting of one regular spiral; but this is seldom the case. It is usually wider on one side than the other, or below than above, as in Fig. 28, where outside the spirals are several loops going partly round the web. The web of Zilla consists entirely of such loops going three-quarters round the web, and

returning, leaving a segment without any
cross-threads, in which is the line from the
centre to the spider's nest, Fig. 29. The spider
is shown carrying a fly to its nest attached to

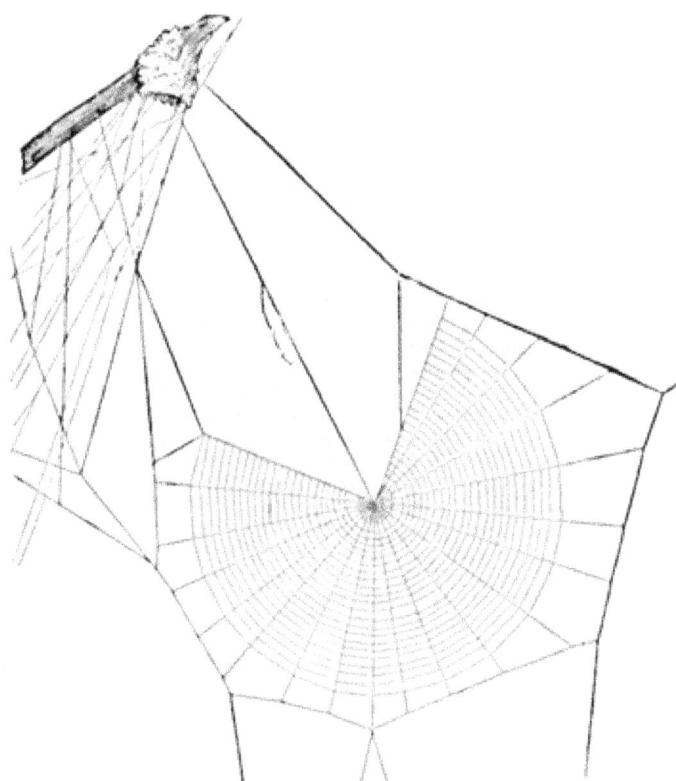

Fig. 29.

the spinnerets; and, if this is its usual habit,
the web with an open segment is certainly more
convenient than a complete one.

The web of *Nephila plumipes*, described by Wilder, consists also of loops running round about quarter of a circle, Fig. 30; and in this web the smooth cross-lines which are first spun are not removed, but remain after it is finished. Fig. 31 shows part of one side of a web; the arrows marking the smooth thread, and the direction in which it was spun.

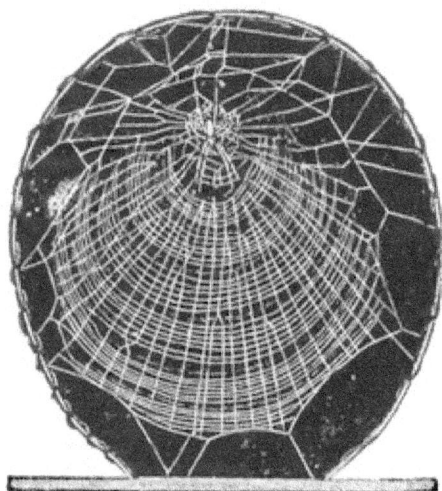

Fig. 30.

Argiope, the large black and yellow autumn spiders, cross the middle of the web with a zig-zag band of white silk, which, as the web is obliquely hung, partly conceals the spider under it. These spiders also spin each side of the web, and two or three inches from it, a screen of irregular threads of unknown use.

The round-web spiders are said to repair

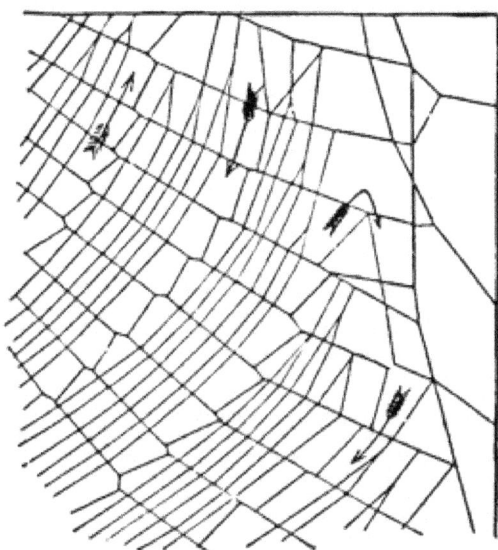

Fig. 31.

their webs by tearing out a dirty, tangled piece, and putting a new one in its place. Wilder says that *Nephila plumipes* tears off and replaces half the web at one time. *Epeira vulgaris* often takes away an old web, and puts a new one in the same place, tearing down the old in pieces, and putting in the rays of the new as it goes along. The spider walks on the nearest sound thread, and gathers in with her front-feet as much old web as she can tear off, and rolls it up with her palpi and mandibles into a ball, and, when it is tight enough not to stick to the web, drops it. As she walks along, gathering up the old web in front, she at the same time spins a new thread behind, and, when she gets to a suitable place, makes it fast as one of the rays of the new web. The common story has it, that the spider eats the old web. She certainly gathers it up in her mouth, and sometimes throws it away at once, but at other times sits and chews it a long time, with apparent pleasure.

Most of the *Epeiridæ* are brightly colored, and make no attempt at concealment when in the web. Others have odd shapes and colors, and hang in the web in such positions that they look like any thing but animals. Some species draw up their legs against their triangular abdomens, and look like bits of bark fallen into the web. Others are long and slender, and when at rest, either in the web or out, lay their legs close together before and behind their bodies, so as to look like straws. Others have oddly shaped abdomens, as Fig. 32, under which the rest of the body is partly concealed.

Epeira caudata, a common gray spider, living in the wood, collects pieces of insects and other rubbish, and arranges it in a line up and down,

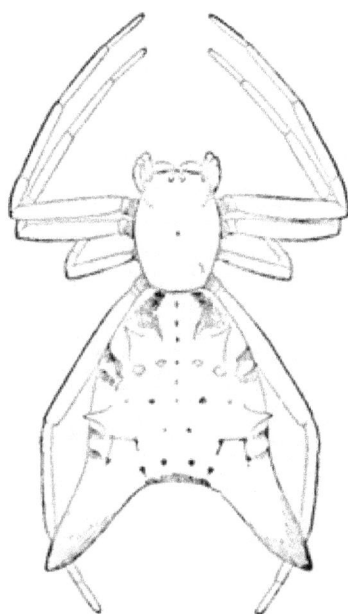

Fig. 32.

across the centre of the web. The spider stands in the centre, and from a short distance can hardly be distinguished from the rubbish. She also hides her cocoons in the web, in the same line of dirt.

The size of the web is usually proportioned to that of the spider; but *Epeira displicata*, which is quarter of an inch long, makes a web only two or three inches in diameter, on the ends of branches of bushes, where it is shaken about, and sometimes blown to pieces, by the wind.

As the spider stands in her web, and feels a slight shake, such as would be caused by a sudden wind, she draws her legs together, pulling the rays tighter, and so making the whole web steady. If, however, the spider is frightened, and has no time to escape, she throws her body back and forth as a man does in a swing, and thus shakes the web so rapidly, that the spider can hardly be seen. The most usual habit, when alarmed, is to drop to the ground, and lie there as if dead.

USE OF SPIDER'S SILK.

Various attempts have been made to use the silk of spiders, and chiefly that of the large round-web spiders, for practical purposes, either by carding the cocoons, or by drawing the thread directly from the spider. The latest experiments and plans for this purpose are those of Professor Wilder in "The Galaxy," vol. viii. He shows how *Nephila plumipes* might be raised in large numbers, each spider kept by herself in a wire ring surrounded by water, fed with flies bred for the purpose from old meat, and milked every day of their thread. Each cocoon of this spider contains from five hundred to a thousand eggs. The young live together for two or three weeks, spin a web in common, and eat one another, or any small insects that come in their way. Then they begin to scatter, and each builds her own web; so that from this time they must be kept separate, or they would eat one another. Every day or two, each spider should be taken down, put into a pair of stocks, and the thread pulled out till it stops coming. In this way Wilder thinks an ounce of thread could be got from each spider during the summer. The thread is from a seven-thousandth to a four-thousandth of an inch thick, and much smoother and more brightly colored, as well as finer, than that of the silk-worm. Several threads would have to be twisted together to get one of manageable

size. The principal difficulties are the space
needed for keeping each spider by herself, and
the amount of labor needed to provide them
with living insects for food, and to draw out
the silk, which would make it too expensive to
use.

CURLED WEBS.

There is a family of spiders called by Blackwall *Ciniflonidæ*, see p. 17, which, besides the

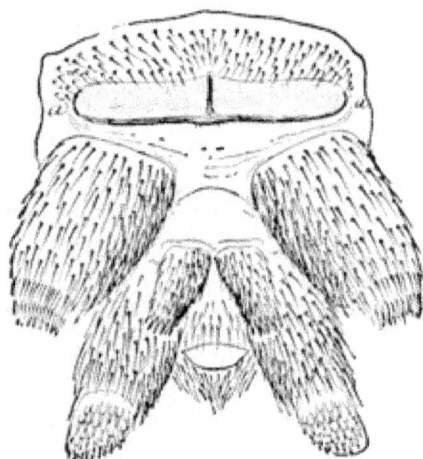

Fig. 33.

usual plain thread, make a peculiar kind of
their own. They have in front of the spinnerets, Fig. 33, an additional spinning-organ
called the cribellum, *a, a.* It is covered with

fine tubes, much finer than those of the spinnerets, set close together.

They also have on the last joint but one of
the hind-legs a comb of stiff hairs, called the
calamistrum, Fig. 34, on the upper side.

Fig. 34.

When they spin their peculiar web, they turn
one of the hind-legs across under the spinnerets, so that the calamistrum is just under
the cribellum, and the foot rests on the opposite leg, Fig. 35. The hind-legs are then

Fig. 35.

moved rapidly back and forth; so that the calamistrum combs out from the spinning-tubes,

and at the same time tangles a band of fine
threads, C. This band is laid along, and
attached here and there to a plain thread, A, B,
so as to make it adhere more readily to an
insect that happens to touch it. As one leg
gets tired, they change, and work with the
other. In the webs of these spiders this adhe-

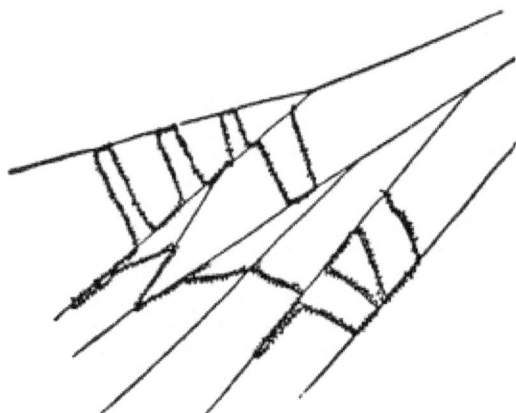

Fig. 36.

sive band can be seen with the naked eye, running about, as in Fig. 36. The webs are usually
irregular, and shaped to fit the place where they
are built, but have, in some part, a tube some-
what like that of the grass spider, Fig. 24,
where the owner hides. Sometimes they are
more or less regular in structure, some of the

threads being parallel, and crossed by shorter ones at regular intervals, Fig. 37. Others are circular, with a tube in the centre which runs into a crack, and from which radiate irregularly the principal threads of the web. Such webs

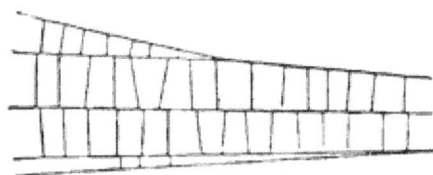

Fig. 37.

are sometimes very numerous on stone buildings, and, as they collect large quantities of dust, seriously disfigure them. The webs alone, when clean, would not be noticed.

THE TRIANGLE SPIDER.

Among those spiders that use the calamistrum is one which makes a web unlike any other. It has been described by Professor Wilder, in the "Popular Science Monthly" for April, 1875, under the name of the "triangle spider." It lives usually among the dead branches around the lower part of pine and spruce trees, and is colored so like the bark, that when it stands, as it usually does, on the

end of a branch, it is easily mistaken for a part
of it.   The web seems to be made in the night.
Wilder saw them early in the morning ; and I,
in the evening, between sunset and dark.   A

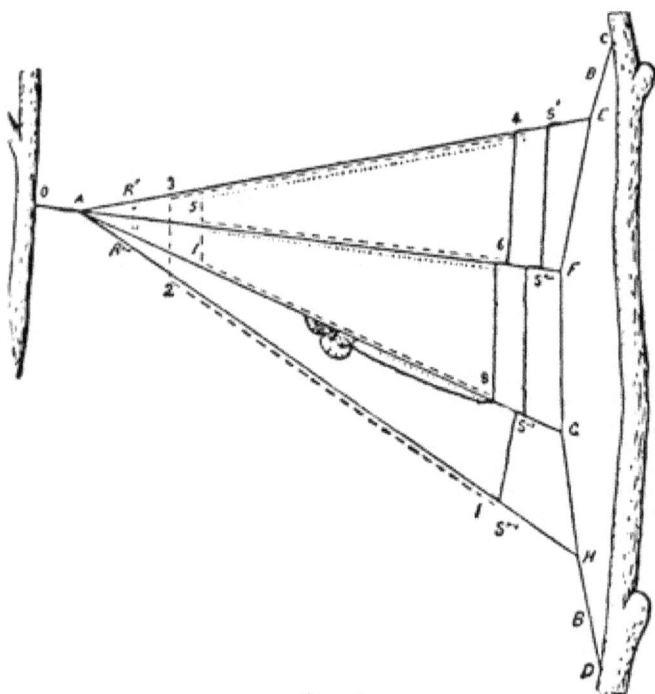

Fig. 38.

single thread five or six inches long runs from
the spider's roost ; and from its extremity radiate four branches, attached to various twigs in
the neighborhood, Fig. 38, AE, AF, AG, AH.

The spider begins to cross them with adhesive threads near the end of the upper ray at S'. After fastening the end of the thread, she walks along toward the centre, scratching away all the time with her calamistrum, till she comes to a place, 5, where she can cross to the next ray. She crosses over, and goes outward toward S", the thread shrinking as she goes, until, when she arrives at S", it is just long enough to reach across to S'. She fastens it by laying it along the ray for a short distance, and goes inward again till she reaches 7, where she crosses to the next ray; and so on till the thread is finished to S''''. Here she stops spinning, and goes up the lower ray to A, and along the upper one to 4, where she starts another cross-thread. This goes on till the whole web is filled, as in Fig. 39, nearly to the centre.

When the web is finished, the spider goes up the thread A *o*, to within an inch or so of the twig to which it is fastened; turns round, and takes hold of the thread with her front-feet; then pulls herself backward with her hind-feet up to the twig. She thus tightens the web, and draws up a loop of thread between her front and hind feet, Fig. 39, lower figure.

The net is now set for use, and she stands
holding it till something touches it ; then she
lets go with her hind-legs, and the net springs
forward, bringing more threads into contact

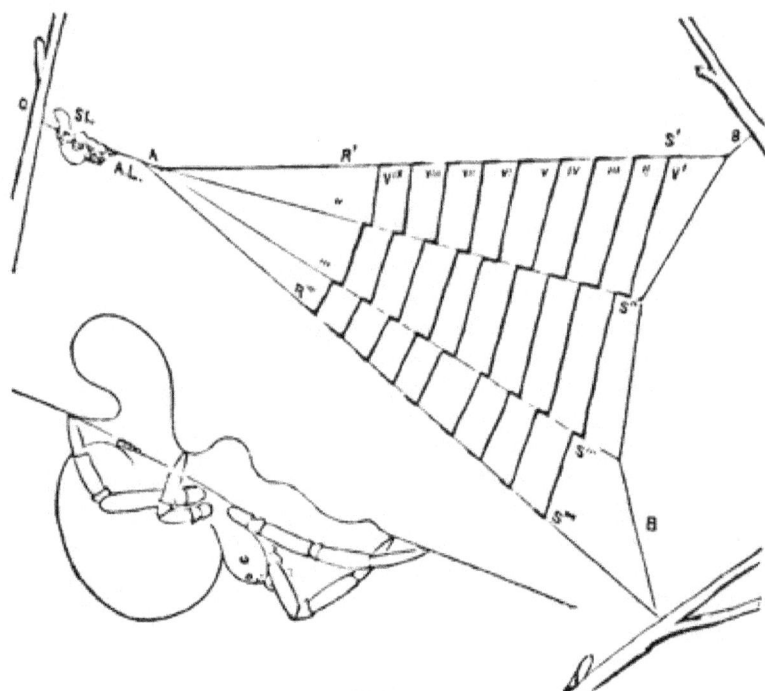

Fig. 39.

with the insect, and sliding the spider along the
line toward A. If she thinks it worth while,
she draws up another loop, and snaps the web
again. When she is satisfied that the insect is

caught, she gathers up part of the web till she comes to him, covers him with silk, and carries him up to her roost.

There are other spiders of this group that make round webs, just like those of the *Epeiridæ*, Fig. 28, except in the adhesive threads being spun with the calamistrum.

FLYING SPIDERS.

Often, in summer, the bushes are covered with threads, attached by one end, blowing out in the wind; and bits of cobweb are blowing about, with occasionally a spider attached. To account for such threads, curious theories have been thought of; among others, that spiders are able to force the thread from their spinnerets, like water from a syringe, in any direction they choose.

If a spider be put on a stick surrounded by water, she manages, in course of time, to get a thread to some object beyond, and to escape by it. To find out how this is done, Mr. Blackwall tried some experiments. He put spiders on sticks in vessels of water, and they ran up and down, unable to escape as long as the air in the room was still. But, if a draught of air passed the spider, she turned her head toward it, and opened her spinnerets in the opposite direction. If the draught continued, a thread was drawn out by it, which at length caught upon something, when the spider drew it tight, and escaped on it. If the air was kept still, or the spider covered with a glass, she remained on the stick till taken off.

These experiments have been repeated, and show that the spider does not shoot or throw the web in any way, but takes advantage of currents of air, and allows threads to be blown out to a considerable distance.

There is a still more curious use of this method of spinning threads ; that is, in flying. Small spiders, especially on fine days in the autumn, get up on the tops of bushes and fences, each apparently anxious to get as high as possible, and there raise themselves up on tiptoe, and turn their bodies up, as in Fig. 40, with their heads toward the wind, and spinnerets open. A thread soon blows out from the spinnerets, and, if the current of air continues, spins out to a length of two or three yards, and then offers enough resistance to the wind to carry the spider away with it up into the air. As soon as she is clear, the spider turns around, and grasps the thread with her feet, and seems to be very comfortable and

Fig. 40.

contented till she strikes against something. Sometimes they rise rapidly, and are soon out of sight; at other times blow along just above the ground.

This habit is not confined to any particular kinds of spiders, but is practised by many small

species of *Erigone*, and by the young of many
spiders of all families, that, when adult, would·
be too large for it. The majority of the spiders
flying in autumn are the young of several
species of *Lycosa*, that seem to spend the
greater part of October and November trying
to get as far above ground as possible. The
best places to watch them are garden-fences in
cities, where they often swarm, and can be more
distinctly seen than on bushes. Large numbers can always be seen, for example, on the
fences around the Common in Boston, every
fine day in autumn, until there has been a long
period of cold weather. Other species fly in
the early part of summer.

Mr. Blackwall observed in Manchester, Eng.,
Oct. 1, 1826, a calm sunny day, that, just before
noon, the fields and hedges were covered over
with cobwebs. So thick were they, that, in
crossing a small pasture, his feet were covered
with them. They had evidently been made in
a very short time, as early in the morning they
were not conspicuous enough to attract his
attention, and the day before could not have
existed at all, as a high wind blew all day. At

the same time large rags of web were floating about in the air, one measuring five feet long, and several inches wide. These appeared to be not formed in the air, but torn from grass and bushes, where they were produced by the tangling of many threads which had been spun separately. They kept rising all the forenoon, and in the afternoon came down again. Not one in twenty had a spider on it. Similar large webs were observed by Lincecum in Texas, and supposed by him to be balloons spun purposely by the spiders.

Mr. Darwin, in the journal of the voyage of "The Beagle," says, that when anchored in the River Plata, sixty miles from shore, he has seen the rigging covered with cobwebs, and the air full of pieces of web floating about. The spiders, however, when they struck the ship, were always hanging from single threads, and never to the floating webs.

A recent account of the signs of weather-changes near the southern coast of the United States mentions as one of them cobwebs in the rigging.

It is still unexplained how the thread starts

from the spinnerets.   It has been often as-
serted that the spider fastens the thread by the
end, and allows a loop to blow out in the wind;
but, in most cases, this is certainly not done,
only one thread being visible.   Sometimes,
while a thread is blown from the hinder spinnerets, another from the front spinnerets is

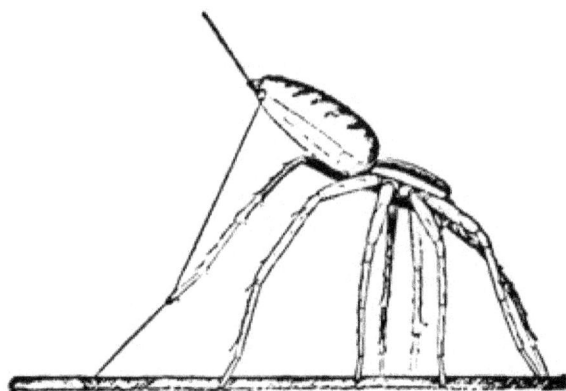

Fig. 41.

kept fast to the ground, Fig. 41 ; so that, when
the spider blows away, it draws out a thread
behind it entirely independent of the one from
which it hangs.

Sometimes, instead of a single thread, several
are blown out at once, like a long brush, as in

Fig. 42, which represents, four times enlarged, an unusually large spider just before blowing off a fence.

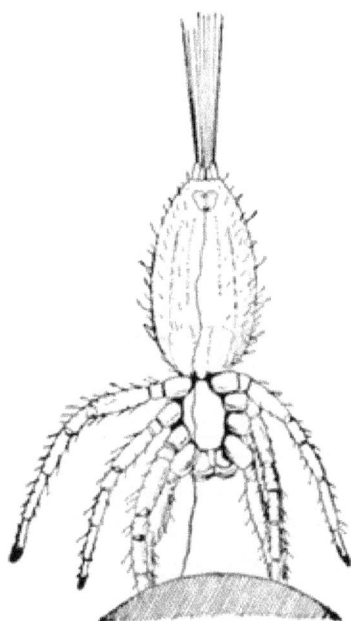

Fig. 42.

CHAPTER IV.

GROWTH OF SPIDERS.

PERSONS unfamiliar with spiders find it hard to tell young from old, and male from female. This is caused, in part, by the great differences between different ages and sexes of the same spider, on account of which they are supposed to belong to distinct species.

The adult males and females are easily distinguished from each other, and from the young, by the complete development of organs peculiar to each sex, which will be described further on.

The males are usually smaller than the females, and have, in proportion to their size, smaller abdomens and longer legs. They are usually darker colored, especially on the head and front part of the body; and markings which

86

are distinct in the female run together and

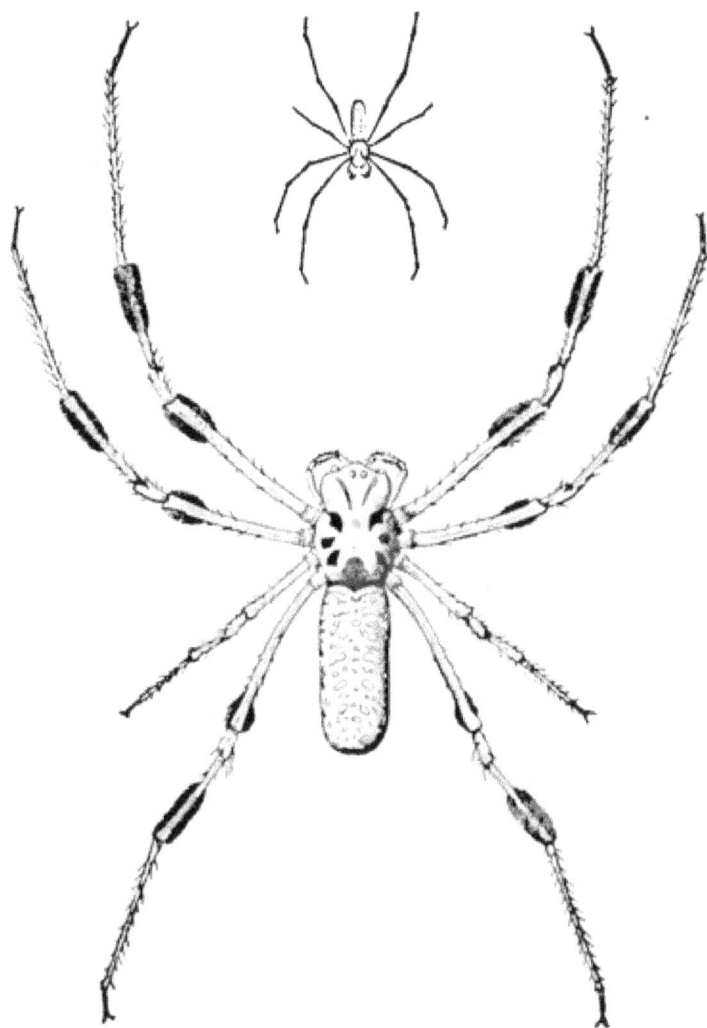

Fig. 43.

become darker in the male. In most species these differences are not great; but in some no one would ever suppose, without other evidence, that the males and females had any relationship to each other. The most extreme cases of this kind are *Argiope* and *Nephila*, where the male is about a tenth as large as the female. Fig. 43 represents male and female of *Nephila plumipes* described by Wilder.

Fig. 44.

The female of one of the common crab spiders is white as milk, with a crimson stripe on each side of the abdomen ; while the male is a little brown-and-yellow spider, with dark markings of a pattern common in the family to which it belongs.

In the genus *Erigone*, which includes the smallest known spiders, the males often have

curious humps and horns on their heads,
Fig. 44. The most extreme example is Fig.
45, where the eyes are carried up on the end of
the horn. The females of all
these species have plain round
heads ; and what use the humps
are to the males nobody knows.

The peculiar organs by which
the adult males and females can

Fig. 45.

always be distinguished are, in the males, the
palpal organs, on the ends of the palpi ; and,
in the females, the epigynum, Fig. 1.

PALPAL ORGANS.

As the male spider gets nearly full grown,
the terminal joints of the palpi become swollen,

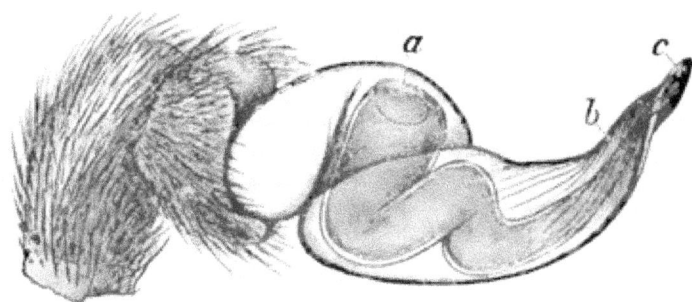

Fig. 46.

and, after the last moult, the palpal organs are
uncovered.

The simplest form of palpal organ is found
in the large *Mygalidæ*, Fig. 6.   It consists of
a hard bulb, Fig. 46, drawn out to a point, in
which is a small hole leading to a sac within.

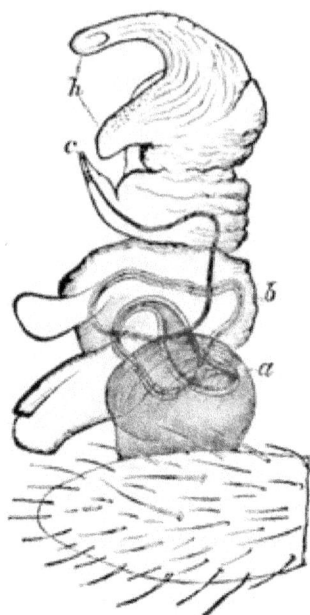

Fig. 47.

In most spiders the terminal joint is flattened, and has a hollow on the under side, in
which the palpal organ is partly concealed.
The bulb is flattened to fit this hollow; and the
point of it is prolonged into a distinct tube of
various shapes, furnished with numerous spines

and appendages. Fig. 47 is the palpal organ of
Epeira vulgaris flattened out, and made trans-
parent. The internal sac is shown at *a ;* and
the tube from it *b* runs round the inside of
the bulb, and ends at the opening *c.* The outside of the organ has various horns and appendages. Fig. 48 is the palpus of another spider,

Fig. 48.

where the outer tube is so long, that it is coiled
up over the basal part of the bulb ; and the end
rests on a strong spine at one side of the palpus.

Not only the terminal joints of the palpi, but
also the next, and sometimes next two joints,
are modified in shape, with the development of

the palpal organ, Fig. 48.   The shape of these organs is very constant in the same species of spider, and thus becomes one of the most convenient marks of such a group.

THE EPIGYNUM.

When the female spider is nearly full grown, there appears a hard, swollen place just in front of the opening of the ovaries, Fig. 1 ; and, after the last moult, the epigynum is uncovered at

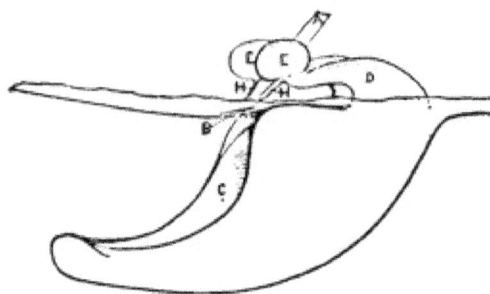

Fig. 49.

this place.   The epigynum, Fig. 49, consists of two spermathecæ, E, which connect by two little tubes, H, H, with the oviduct near its mouth, and by two larger tubes, D, with the outside of the spider.   The mouths of these larger tubes are often surrounded by various

hard appendages to support and guide the palpal organs when the latter are thrust into them.
These parts, like the palpal organs, furnish convenient marks for distinguishing species. The
spermathecæ, E, vary but little in shape in
different spiders ; but the tubes H and D are
often lengthened, and twisted into shapes nearly as complicated as those of the palpal organs.

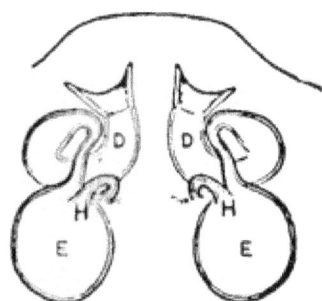

Fig. 50.

Fig. 50 is the epigynum of a small *Theridion*,
where the arrangement of parts can be very
distinctly seen. E, E, are the spermathecæ ;
H, H, the tubes opening into the oviduct ; and
D, D, the tubes opening outward. Fig. 51 is
the epigynum of another closely allied species,
where the tubes D, D, are very much elongated
and twisted up, corresponding to the long dis-

charge-tube of the palpal organ of the male of
the same spider, Fig. 48.

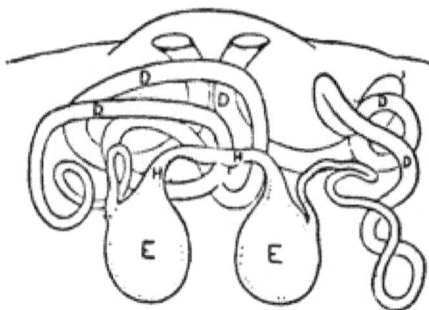

Fig. 51.

USE OF THE PALPAL ORGANS AND EPIGYNUM.

When the reproductive cells of the male
spider are mature, he discharges the liquid containing them on a little web spun for the purpose ; dips his palpal organs into it, and in a
few moments takes up the whole, it is supposed, into the little sacs, Figs. 47, 48, inside
the bulb ; then he seeks the female, and inserts
the palpal organs into her epigynum. The soft
part at the base of the organ swells up, and
presses in the discharge-tube, and probably
forces out the contents of the bulb into the
spermathecæ, E, E, from which it escapes, in

course of time, by the tubes, H, H, into the oviduct, and fertilizes the eggs about the time they are laid.

One palpal organ is usually inserted at a time, and, after a while, taken out, and replaced by the other; this change being repeated many times by the same spider. Among the *Lycosidæ*, Fig. 10, the male leaps on the back of the female, and is carried about by her, Fig.

Fig. 52.

52. He reaches down at the side of her abdomen, and inserts his palpi in the epigynum underneath. In *Linyphia* and *Theridion* the male and female live peaceably together for a long time in the same web. The male reaches from in front under the female, Fig. 53, and inserts his palpal organs, one after the other, for hours together. In *Agalena* the male is the stronger of the two sexes. He takes the

female in his mandibles, and lays her on one

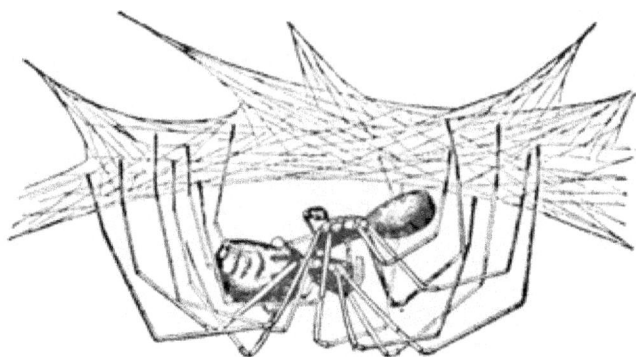

Fig. 53.

side, Fig. 54, and inserts one of his palpi.
After a time, he rises on tiptoe, turns her

Fig. 54.

around and over, so that she lies on the other
side, with her head in the opposite direction,

and inserts the other palpus.
The female lies as though
dead. In *Nephila* and *Argiope*, where the male is very
small, he stands on the upper
edge of the web while the
female is in her usual position in the centre. After
feeling the web with his feet
for some time, he runs down
to the centre so lightly as not
to disturb the female, and
climbs about over her body
for some minutes, in an apparently aimless way. She
takes no notice of him at
first ; but at length, especially if he approach the under
side of her abdomen, she
turns, and snaps at him with
her jaws. He is usually
nimble enough to dodge between her legs, and drop out
of the web, and, after a while,
climbs up to the top, and

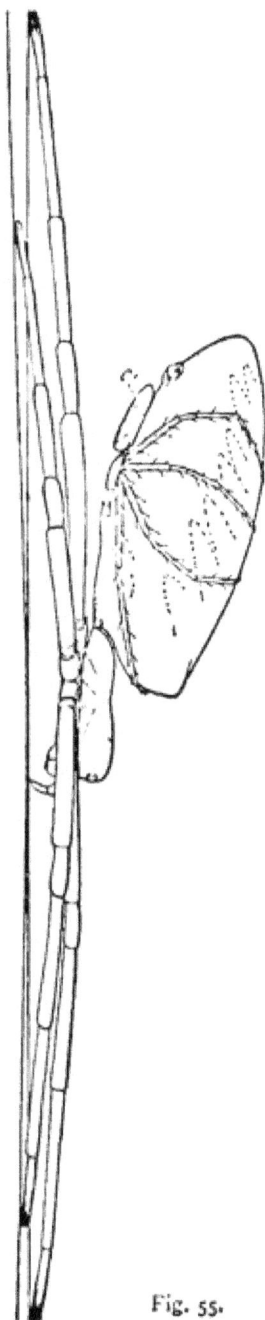

Fig. 55.

begins over again. In these encounters the males are often injured; they frequently lose some of their legs: and I have seen one, that had only four out of his eight left, still standing up to his work.

At length the male succeeds in getting under the female's abdomen, and inserting his palpi into the epigynum. Fig. 55 shows the female hanging in the web, with the male at *a*, with his legs grasped around her abdomen.

The habits of these spiders furnish the grounds for the popular story, that female spiders regularly eat the males. No doubt it occasionally happens, where the female is the larger of the two; but in many species they live together for some time in the same web, or in a nest spun for the purpose; in some cases, before the female has reached the adult state.

LAYING EGGS.

When the eggs are mature, the female proceeds, like the male, to make a little web, and lays the eggs on it. Then she covers them over with silk, forming a cocoon, in which the young remain till some time after they are hatched. The laying of the eggs is seldom seen ; for the spider does it in the night, or in retired places ; and often, in confinement, refuses to lay at all.

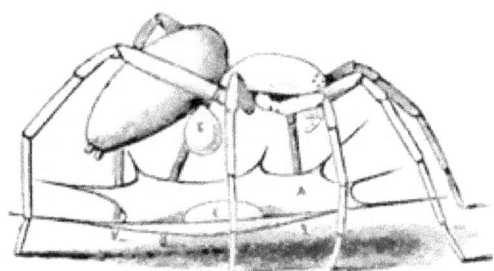

Fig. 56.

The female *Drassus*, Fig. 56, spins a little web A across her nest, and drops the eggs E on it, as in the figure. They are soft, and mixed with liquid, and are discharged in one or two drops, like jelly. They quickly soak up the liquid, and become dry on the surface, sometimes adhering slightly together.

After the eggs are laid, the spider covers them with silk, drawing the threads over them from one side to the other, and fastening them to the edges of the web below. When the covering is complete, she bites off the threads that hold the cocoon to the nest, and finishes off the edges with her jaws.

The *Lycosidæ* make their cocoons in the same way, but rounder, and showing only slightly the seam where the upper part was attached to the lower.

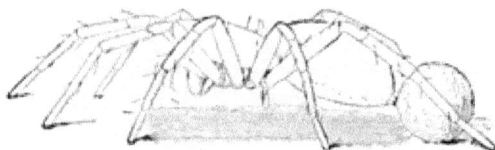

Fig. 57.

The *Lycosas* carry their cocoons about, attached to the spinnerets, as in Fig. 57, bumping them over the stones without injury to the young inside.

Many spiders make their cocoons against a flat surface, where they remain attached by one side. *Attus mystaceus* spins, before laying, a thick nest of white silk on the under side of a stone. In this she thickens a circular patch on the upper side, next the stone, and discharges her eggs upward against it, Fig. 58. They adhere, and are covered with white silk. I once had a spider of this species lay her eggs,

in confinement, in a nest the under side of
which had been cut away. Instead of com-
pleting the cocoon properly, she ate the eggs

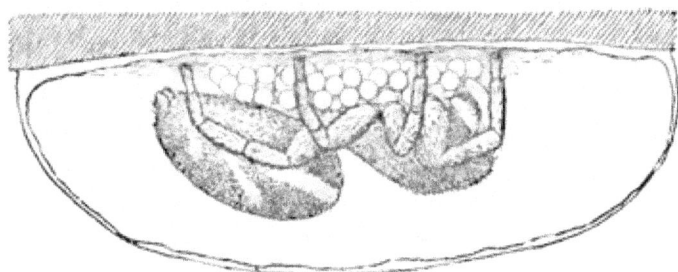

Fig. 58.

immediately after laying. *Epeira strix* spins,
before laying, a bunch of loose silk, Fig. 59.
She touches her spinnerets, as in the figure on

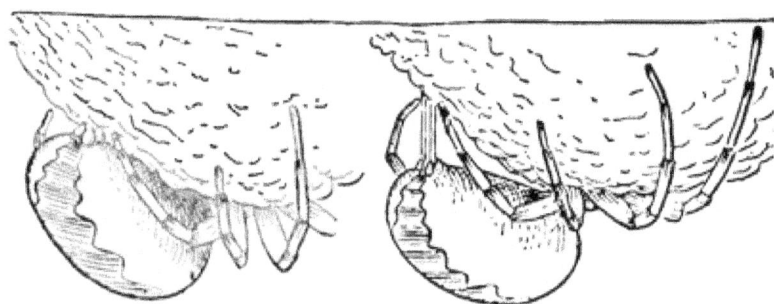

Fig. 59.

the left, draws them away a short distance, at
the same time pressing upward with the hind-
feet, as in the figure on the right ; then moves

the abdomen a little sidewise, and attaches the band of threads so as to form a loop. She keeps making these loops, turning round, at the same time, so as to form a rounded bunch of them, into the middle of which she afterwards lays the eggs, as in Fig. 60. The eggs, which are like a drop of jelly, are held up by the loose threads till the spider has time to spin under

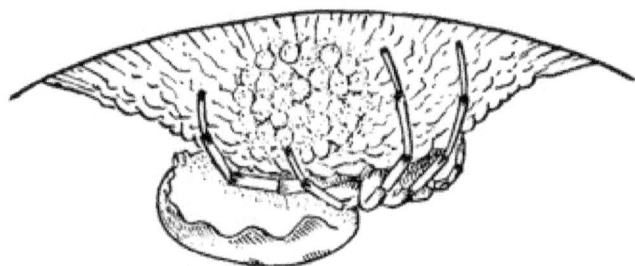

Fig. 60.

them a covering of stronger silk. *Epeira vulgaris* makes a similar cocoon upward, downward, or sidewise, as may be most convenient.

Most of the *Theridiidæ* make cocoons of loose silk, held up in the web by numerous threads. Some hang the cocoon by a stem, Fig. 61.

The large species of *Argiope* makes a big

pear-shaped cocoon hanging in grass or bushes, Fig. 62. A stem of loose brown silk is first made, and under this the eggs attached (at any rate this had been done in one which had been

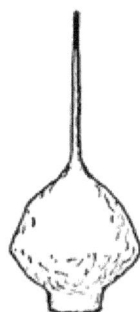

Fig. 61.

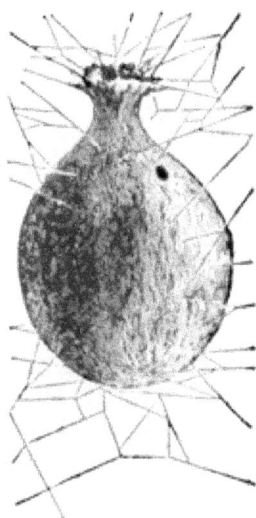

Fig. 62.

abandoned unfinished) ; then a cup-shaped piece is made under the eggs ; the bunch of loose silk is spun over all, and finally the paper-like shell.

ESCAPE FROM THE COCOON.

These cocoons of *Argiope* are made late in the summer, and the young stay in them till the next season. Out of six hundred cocoons collected by Wilder in the spring, less than a quarter were entire, the rest being pierced, or torn in some way, by birds or insects; so that the spiders were saved the trouble of gnawing their way out, as they can if obliged to.

I once noticed a small *Theridion* gnawing at its soft cocoon, and found that one side had been made in this way much thinner than the rest of the cocoon. I put her, with the cocoon, in a bottle where I could watch her; and she soon commenced biting again, and kept it up the rest of the day. The following night the young came out.

Many spiders remain by their cocoons till the young come out; but other species, making similar ones, go away, or die, and the young get out themselves when they are old enough.

The young of *Micaria* cut a smooth round hole in their paper-like cocoon, just large enough for them to come out one by one.

PARASITES.

The eggs in the cocoon are very liable to be eaten by parasitic insects. Certain wingless *Hymenoptera* are always hunting around in the

neighborhood of spiders' nests, and may sometimes be seen trying to stick their ovipositor
through a cocoon. If they succeed, their eggs
hatch before the spiders, and eat the latter up.
Other parasites lay eggs on the backs of young
spiders, and the larva lives attached to the outside till it gets nearly as large as the spider
itself.

GROWTH IN THE EGG.

The egg of a spider, like that of any other
animal, is a cell which separates from the body
of the female, and afterwards unites with one
or more cells which have separated from the
body of the male. This fertilization of the
eggs probably takes place when they have
reached their full size, and are about to be laid.

After the eggs are laid and hardened, it is
very easy to watch their development. They
grow just as well anywhere else as in the
cocoon, and, in order to see through the shell,
it is only necessary to cover the egg to be
examined, with oil, alcohol, or any liquid that
will wet it.

Just after it is laid, the egg looks like Fig.
63, *a;* or, if the egg is more opaque, only the

ends of the lobes can be seen like irregular
lumps. The first sign of growth is the divis-
ion into two, Fig. 63, *b.* These divide into

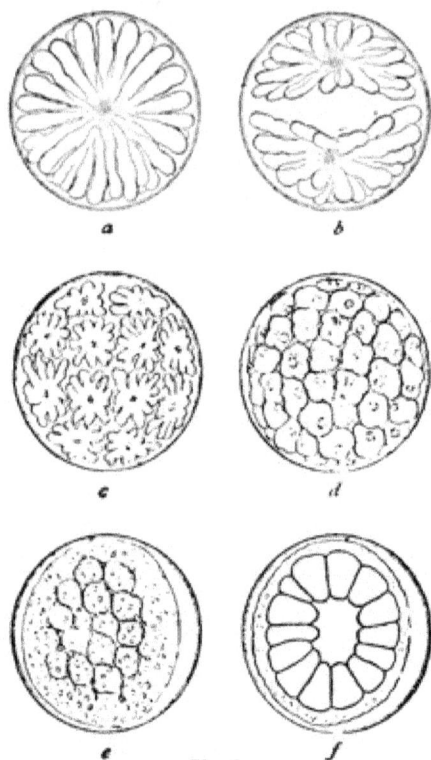

Fig. 63.

four, into eight, and so on, Fig. 63, *c, d.* At
first the divisions are all alike; but at length
they divide into two kinds, — small ones, with a

dark spot in the middle, which cover the outside of the egg ; and larger ones that occupy the
inside. Fig. 63, *e*, shows an egg at this stage,
where the large inner cells show through the
layer of outer ones. Fig. 63, *f* is a section of
the same egg. The stages shown in *b* and *c*
are seldom clearly seen, because the divisions
are crowded together and too opaque ; but *d* and
e can be watched in any common spider's eggs.
The rate of growth varies according to circum-
stances. Some eggs laid in autumn develop
slowly all winter, while others laid in summer
are ready to hatch in a fortnight.

In the eggs of the long-legged cellar spider,
laid in June, in about four or five days the young
spider becomes lengthened out into a sort of
barrel shape ; and six whitish rings run half way
round it, on each of which appears soon after a
pair of little knobs, one each side, Fig 64, *a*.
These are the six segments of the thorax, and
the six pairs of limbs ; and their gradual growth
is shown in Fig. 64, *b, c, d*. In *a* there is no sign
of a head or abdomen, except the more opaque
ends of the embryo ; but shortly after there
appears an opaque knob at one end, Fig. 64, *b*,

under which is a pair of little knobs, such as appeared at first on the thoracic segments; then appear two pairs, then three, and so on, till there are six pairs, which mark the six segments of the abdomen.   Up to this time, the embryo has

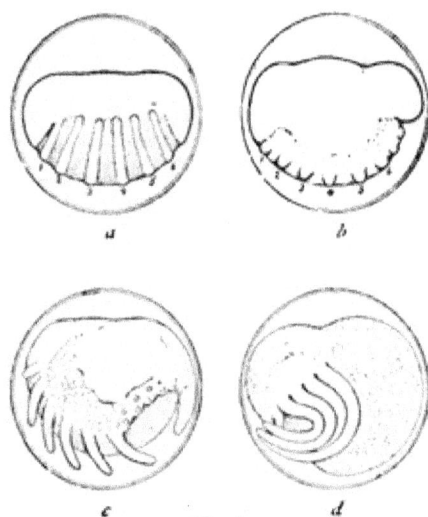

Fig. 64.

been rolled up with the under side outward; but now it begins to turn, and in a day or two has its back outward, Fig. 64, *c*.   The constriction between the thorax and abdomen begins about this time; and in a few days more the spider is ready to hatch, Fig. 64, *d*.

YOUNG SPIDERS.

The hatching occupies a day or two. The shell, or rather skin, cracks along the lines between the legs, and comes off in rags ; and the spider slowly stretches itself, and creeps about. It is now pale and soft, and without any hairs

Fig. 65.

or spines, and only small claws on its feet ; but, in a few days, it gets rid of another skin, and now begins to look like a spider. The eyes become darker colored ; marks on the thorax become more distinct, and a dark stripe appears across the edge of each segment of the abdomen. The hairs are long, and few in number, and arranged in rows across the abdomen and along the middle of the thorax, Fig. 65. Before the next moult, they usually leave the cocoon, and for a time live together in a web spun in common. A brood of young *Epeira* may often be seen looking like a ball of wool in the top of a bush, while below them, connected by threads to their roost, are the skins left at their second moult, and farther down, also connected by threads, the cocoon with the first skins.

Dolomedes spins a nest in which the young live for a while after hatching.

The young of the running spiders, *Lycosidæ*, when they come out of the cocoon get on their mother's back, and are carried round by her for some time.

Where large broods of young spiders live together, they soon begin to eat one another; and, if kept in confinement, one or two out of a cocoon full, may be raised without any other food.

Wilder noticed this in *Nephila plumipes*, and believes it is the natural habit of young spiders, and not the result of confinement.

As spiders grow larger, they have to moult

from time to time. This process is shown
by Wilder in Figs. 66, 67; and I have seen
the same operation in *Argiope*. The spider

Fig. 66.

Fig. 67.

hangs herself by a thread from the spinnerets
to the centre of the web. The skin cracks
around the thorax, just over the first joints of
the legs; and the top part falls forward, being

held only at the front edge. The skin of the abdomen breaks irregularly along the sides and back, and shrinks together in a bunch. The spider now hangs by a short thread from the spinnerets, and works to free her legs from the old skin, Fig. 66. This takes about quarter of an hour; and then she drops down, hanging by her spinnerets like a wet rag, Fig. 67.

If struck while in this condition, she can do nothing, not even draw her legs away. After ten or fifteen minutes, the legs begin to strengthen; and she draws them gradually up toward her, works them up and down a few times, and is soon able to get into the web again.

Blackwall observed nine moults in *Tegenaria civilis*, a spider that lives several years. Many species, and among them some of the largest, live only one year, hatching in the winter, leaving the cocoon in early summer, and laying eggs and dying in autumn. Other species seem to require two years for their growth; hatching in summer, passing their first winter half grown, growing up the next summer, but laying no eggs till the second spring. Some species are found adult at all seasons, and may live several years.

After spiders have passed their second moult, they usually live in the same places, and follow the same habits, as the adults.

The running spiders live usually on the ground, often near water, but some kinds in the hottest and dryest places. A few species live near water, and are accustomed to run about on its surface, without becoming wet. The *Theridiidæ* almost all live in the shade, and always upside down in their webs. Some species live always in caves; and one in the deepest part of the Mammoth Cave has no eyes. Some spiders live only on high mountains, never appearing below the tree line. Some species seem to prefer certain kinds of plants. The horizontal branches of spruces, for instance, are particularly convenient for the webs of some species of *Theridion*. The water-spider, that builds its nest and lives on water-plants, has been already mentioned, and also the *Argyrodes*, that makes its home in the webs of other spiders. During winter immense numbers of spiders that have spent the summer under stones, in webs, and on plants, hide away among fallen leaves, and there live through the coldest and wettest weather, ready to move on the first warm day. During a thaw they often come out on the snow in great numbers.

Several house spiders have probably been imported, like rats, and are found all over the world; while other most common species never spread beyond the countries where they are most abundant.

BOOKS ABOUT SPIDERS.

CLASSIFICATION. — Thorell's "Genera of European Spiders," in "Acta Regiæ Societatis Scientiarum Upsalensis," 1869, and Thorell's "Synonymes of European Spiders," contain a complete history of the classification of the spiders of Northern Europe, with references to all the descriptions of genera and species, and remarks on the use of names and groups by different authors. The great resemblance between the European and North-American spider faunæ make these the most useful books for American students. Simon's "Arachnides de France," a work not yet completed, describes all the spiders in France, and refers to descriptions of the other European species. It contains tables by which the genus and species to which any spider belongs can be found by the use of a few prominent characters.

ANATOMY. — Siebold's "Anatomy of the Invertebrata" contains a good general account. Bertkau describes, in "Traschel's Archiv für Naturgeschichte," the mandibles in 1870, the respiratory-organs in 1872, and the sexual-organs in 1875. Oeffinger describes the spinning-glands in "Archiv für Microscopische Anatomie," 1860.

EMBRYOLOGY. — Claparède, Utrecht, 1862, and Balbiani, in "Annales des Sciences Naturelles," 1872, describe the growth of the egg from segmentation to

hatching. H. Ludwig, in "Zeitschrift für Wissenschaft-
liche Zoölogie," 1876, gives an account of the segmentation in eggs of Philodromus.

HABITS. — Walckenaer's "Histoire Naturelle des Ap-
teres" goes over the whole subject. Blackwall, in
"Researches in Zoölogy," 1834, describes the web-
making of Epeira, and the flying habits of spiders.
Blackwall also writes on habits in "The Spiders of
Great Britain and Ireland," published by the Ray
Society, 1864, and in various papers in "Transactions
of the Linnæan Society," 1833 to 1841. Menge's
Lebensweise der Arachniden in "Schriften der Natur-
forchenden Gesellschaft in Danzig," 1843, goes over
the whole subject, and is particularly useful on the sex-
ual habits. The same author continues the subject in
"Preussische Spinnen," published by the same society,
beginning in 1866, and not yet finished. The habits of
the water-spider are described by Mr. Bell in "Journal
of the Linnæan Society," 1857. The trap-door spiders
and their habits are described by J. T. Moggridge in
"Harvesting Ants and Trap-door Spiders," published by
L. Reeve & Co., London, 1873, and Supplement, 1874.
Prof. B. G. Wilder has published several papers on the
habits of American spiders, the most useful of which
are the following: on Nephila plumipes from South Car-
olina, "Proceedings of the Boston Society of Natural
History," 1865; Practical Use of Spider's Silk in "The
Galaxy," July, 1869; Habits of Epeira riparia, Moulting
of Nephila plumipes, and Nests of Epeira, Nephila, and
Hyptiotes, in "Proceedings of American Association for
Advancement of Science," 1873; the Triangle Spider in
"Popular Science Monthly," 1875.

INDEX.

———◆———

Electrotyped by C. J. Peters & Son, Boston, Mass.

PUBLICATIONS

OF THE

NATURALISTS' AGENCY.

————◆◆◆————

ABBOTT'S Stone Age. Stone Age in New Jersey. By Dr. C. C. Abbott. 8vo. pp. 136. 60 plates. Paper $1.00.

ABOUT Insects and How to Observe Them : an elementary treatise on the Structure and Classification of Insects, by A. J. Ebell, Ph. B. 12mo. Pamphlet. 63 wood-cuts. 30 cents.

BIRDS of North America. By Spencer F. Baird, John Cassin, and George N. Lawrence. With an Atlas of One Hundred Elegant Plates. 2 vols., 4to. Cloth, $15.00. Plates, separate, $10.00. Text, separate, $5.00. A few sets with the plates beautifully colored by hand, at $20.00.

The present work is, in part, a reprint of the General Report on North American Birds presented to the Department of War, and published in October, 1858, as one of the series of " Reports of Explorations and Surveys of a Railroad Route to the Pacific Ocean." In these volumes, however, will be found many important additions and corrections. The Atlas contains 100 plates, representing 148 new or unfigured species of North-American birds. Of these plates, about 50 appear for the first time, having been prepared expressly for this work. The remainder form the ornithological illustrations of the Reports of the Pacific Railroad and United States and Mexican Boundary Surveys, and are distributed throughout the numerous volumes composing those series. All have, however, been carefully retouched and lettered for this edition and quite a number redrawn entirely from better and more characteristic specimens. In fact the plates have been prepared expressly for the present edition, with the utmost care, and embrace one hundred species of birds not figured by Audubon.

In the volume of text will be found a complete account of all the Birds of North America, brought down to the present time, including accurate descriptions of all known species: their geographical distribution: and, as far as possible, all other information necessary to a complete summary or manual of North-American ornithology. Extended bibliographical notices, embracing full reference to very nearly all authors on American ornithology, have been added, and will be found to be of high interest to the student and naturalist.

CHECK List of Coleoptera.   Check List of
Coleoptera of America, North of Mexico. By G. R. Crotch,
M. A. Svo. Paper, $1.00.

COUES' Key to North American Birds.   Key
to North American Birds, containing a concise account of
every species of Living and Fossil Birds at present known
from the Continent north of the Mexican and United States boundary. By Dr. Elliott Coues. Illustrated by 6 steel plates and
upwards of 250 wood cuts. 4to. Cloth, $7.00.

" It is a book of inestimable value to the naturalist, and should
be found in the library of every such person in the land."—*Boston
Traveller.*

" We have no hand-book of similar character, and none that
occupies the place it completely fills."—*Golden Age.*

" A thorough and reliable treatise, comprehending the entire
subject."—*American Sportsman.*

" We can cheerfully recommend it to those who wish a reliable
manual of the birds of North America, in a sufficiently portable
form for ready reference."—*The Independent.*

" The only general exposition of this department of American
ornithology that has yet been made."—*Atlantic Monthly.*

" There is a freshness and boldness in the manner in which facts
are handled, which will be extremely acceptable."—*Nature.*

" No expense has been spared in the preparation of this volume.
The woodcuts are so well executed they would easily pass for some-
thing better. The index is complete; so is the glossary."—*Chicago
Times.*

" The descriptions are exceedingly complete and minute; the
large number of illustrations serve to make the text more clearly
understood, and the volume is a very valuable contribution to ornithology."—*Boston Journal.*

" With the help of this 'Key' the veriest tyro can, with very
little trouble, identify his specimens, and obtain a knowledge and
understanding of American birds, impossible to be found in any
other work."—*Army and Navy Journal.*

" This book will be welcomed both by the amateur and profess-
ional ornithologist as a valuable contribution to the list of books
treating of North American birds. While Dr Coues has modestly
called his work a 'Key,' it is in reality much more than the title in-
dicates. The typographical execution of the work is in every way
worthy of it, and the cuts are very clear and instructive."—*The Nation.*

" A more elegant scientific publication than the 'Key to North
American Birds,' just issued by this house, is not to be found.
This work, of which Elliott Coues, M. D., is the author, forms a
very valuable and exhaustive treatise upon the birds of the continent north of Mexico. The large number of plates and of wood-
cuts, renders it especially interesting, and the style of its publication is almost sumptuous."—*Boston Post.*

BIRDS of the Northwest. A Hand-Book of American Ornithology, containing Accounts of all the Birds Inhabiting the Great Missouri Valley, and many others: together representing a large majority of the Birds of North America, with copious Biographical details from personal observation, and an extensive Synonomy. By ELLIOTT COUES, M. D. Svo. 791 pages. Cloth, $4.50.

COUES' Field Ornithology. Field Ornithology, Comprising a Manual of Instruction on Collecting, Preparing and Preserving Birds, with which is issued a Check List of North American birds. By ELLIOTT COUES, M. D. Svo. Cloth, $2.50.

This manual takes the student into the field, and tells him all about practical ornithology, how to shoot, skin and stuff birds; how to equip himself most conveniently and effectively; how to acquire woodcraft and qualify himself to be a good working Ornithologist. Avoiding all technicalities, it tells him, in familiar style, all he wants to know about the matter; it is full of the most useful, practical directions and suggestions for field work, covering the whole ground, and treating of a wider range of topics than have before been presented in this connection. The portion on TAXIDERMY is especially minute and detailed in its accounts of the manipulations which enable one to make the greatest number of skins, in the best manner and shortest time; how to pack, transport and preserve them, and how to take care of the cabinet. With this work in hand no one need hesitate to undertake the business of collecting for want of knowing how to begin, nor for fear of being unsuccessful. It also treats fully of the collection of Nests and Eggs, and of various collateral matters. The student will find himself taught everything he need know from the cleansing of a soiled feather or mending of a broken egg, up to the general qualifications for success.

COUES' Check List of Birds. A Check List of North American Birds. By ELLIOTT COUES, M. D. The check list gives the common and scientific names of all species and varieties of birds, known to inhabit North America, arranged in a generally accepted sequence, named in accordance with the " Key," and so printed on only one side of the page, as to be susceptable of use in neatly labelling collections. Svo. 75 cts.

FLOWER Object Lessons. Or First Lessons in Botany. A familiar description of a few flowers. From the French of M. EMM. LE MAOUT. Translated by Miss A. L. PAGE. 16mo. cloth, .50.

This little work of fifty-five pages, illustrated by forty-seven woodcuts, has been translated for the purpose of placing within the reach of those interested in object teaching a manual that is most admirably adapted for the purpose, and is offered to parents and teachers with the belief that it fully supplies a gap in the literature of our country.

GENTRYS' Life-Histories of Birds. Life-His-

tories of the Birds of Eastern Pennsylvania. By THOMAS G. GENTRY. 2 vols., 12mo. cloth, $4.00.

"Upon careful examination of Mr. Gentry's 'Life-Histories of Birds,' I can pronounce it to be a very original work, full of minute and precise observations upon points too often overlooked, and giving much information not to be found elsewhere. I welcome it as a valuable addition to the literature of my favorite branch of science."—*Dr. Elliott Coues, Smithsonian Institution, Washington, D. C.*

"His comments upon the architecture of birds, and his minute lists of their fare, are especially new, valuable, and interesting. It is rare that a book with a so thoroughly out-of-door odor comes to us, and we predict for it a cordial welcome."—*Forest and Stream.*

"I consider your 'Life-Histories of Birds of Eastern Pennsylvania' an exceedingly interesting work, and one valuable to the ornithologist, as well as entertaining to the general reader. The 'bill of fare' given in the account of each species is particularly valuable."—*Prof. Robert Ridgway, Smithsonian Institution, Washington, D. C.*

"I shall look forward with great interest to the conclusion of the work"—*Prof. J. A. Allen, Museum of Comparative Zoology, Cambridge, Mass.*

"The considerable number of new observations included in it, give it a permanent value."—*Prof. E. D. Cope.*

"It is an interesting work, and I hope you will continue it."—*Prof. O. C. Marsh, Yale College.*

"I am glad you have given so succinct an account of your observations. I prize the volume very highly."—*Prof. N. H. Winchell, Minneapolis, Minn.*

"The author's statements, for the most part, are based on his own experience, and possess the merit both of precision and freshness. The volume is written in an attractive style, with great fulness of description, though free from burdensome details."—*N. Y. Tribune.*

"The comprehensiveness of the work is wholly praiseworthy. An Appendix and Indices, which partly enhance the value of the work, are also furnished. To a farmer's household, for a student of ornithology and oology, for all who give attention to or have interest in the birds of this part of America, the work has no equal so readily obtainable."—*Boston Traveller.*

MAMMOTH Cave and its Inhabitants. The

Mammoth Cave and its Inhabitants, or Descriptions of the Fishes, Insects, and Crustaceans found in the Cave, with figures of the various species, and an account of allied forms, comprising notes upon their Structure, Development and Habits. with remarks upon subterranean life in general. By A. S. PACKARD, JR. and F. W. PUTNAM. 8vo; Plates and Cuts. Cloth, $1.25.

INVERTEBRATES of Vineyard Sound. Report

upon the Invertebrate Animals of Vineyard Sound and Adjacent Waters; with an Account of the Physical Features of the Region, by Professors A. E. VERRILL and S. I. SMITH. 478 pages 8vo, with 38 plates Paper, $3.00.*

This forms a very convenient and valuable Manual of the Marine Zoology of the Atlantic Coast. It contains:

1st. Popular descriptions of the animals of the entire southern coasts of New England and New York, their habits, places of occurrence, etc., with detailed accounts of the oysters, clams, crabs, lobsters, etc., etc., of various kinds directly useful or injurious to man;

2d. Lists of the species found in the stomachs of fishes, showing the food of our native fishes;

3rd. Habits and metamorphoses of the lobsters and other crustacea;

4th. Systematic catalogue of the invertebrates of southern New England and adjacent waters, their synonymy, geographical and geological distribution, with detailed descriptions of large numbers of new species; and a copious Index and Table of Contents. The number of species enumerated is 725 The plates include about 300 excellent figures.

MINOT'S Birds of New England. Land-

Birds and Game-Birds of New England, with Descriptions of Birds, their Nests and Eggs, their Habits and Mates. By H. D. MINOT. Illustrated by outline cuts. 456 pages, 8vo. Cloth, $3.00.

" The present treatise puts Mr. Minot foremost among the ' local' writers on Ornithology in this country, and fairly in line with the fewer ornithologists, whose works are citable as authoritative on the general subject."—*Nation.*

" Its merits entitle it to full recognition by ornithologists, while they commend it very highly to the student and amateur."—*Nuttall Bulletin.*

" The author tells everything that is known about the birds of which he writes, the simple and obvious, as well as the more abstruse, not assuming, as too many writers upon such subjects do, that his readers are already familiar with the elementary text-books of science."—*N. Y. Evening Post.*

" The newest student need not fear that Mr. Minot's book will be beyond him, and need not hesitate to trust it implicitly; while the advanced scholar or worker will find a store of information nowhere more conveniently accessible. This is a work, which, in short, we can confidently recommend to our readers as one with which they cannot fail to be pleased. It probably gives more for the money than any one now before the public."—*Forest and Stream.*

" The land-birds and principal game-birds of New England have received more satisfactory treatment, than has been given them in any previous special treatise on the subject. It affords a fund of fresh information and a positive advance in our knowledge of the habits of New England birds."—*Dr. Elliott Coues.*

MONOGRAPH of the Geometrid Moths, or
Phalænidæ of the U. S. By A. S. PACKARD, JR. 4to., pp. 607. 13 Plates. $6.00. This is the only work on the Geometrid Moths published in America, and is the most complete monograph which has yet appeared, every species is figured, and when they are known, the larva and pupa. In addition to these there are nearly 200 anatomical drawings, illustrating the venation, development, etc. This work should be in the hands of every Entomologist.

PACKARD'S Guide to the Study of Insects.
A Guide to the Study of Insects, and a Treatise on those Injurious and Beneficial to Crops, for the use of Colleges, Farm-Schools, and Agriculturists. By A. S. PACKARD, Jr., M. D. With 15 plates and 670 wood cuts. $5.00.

The guide has already been introduced either as a text-book or for reference in Harvard University, Williams College, Dartmouth College, Antioch College, Mass. Agricultural College, and Oxford and Cambridge Universities, England.

Certainly the best manual of Entomology which the English reader can at present obtain.—*Nature, London.*

PACKARD'S Half-Hours with Insects. Half
HOURS with Insects. A Popular Account of their Habits, Modes of Life, &c.; which are beneficial and which are injurious to vegetation. By A. S. PACKARD, JR., of the Peabody Academy of Science. Colored plate, 260 wood-cut illustrations. Crown 8vo., Cloth. $2.50.

The subjects treated are: Insects of the Garden, Relation of Insects to Man, Insects of the Plant House, Edible Insects, Insects of the Pond and Stream, the Population of an Apple Tree, Insects of the Field, Insects of the Forest, Insects as Mimics, Insects as Architects, Social Life of Insects and Mental Powers of Insects.

PACKARD'S Common Insects. Our Common Insects.
A Popular Account of the more common Insects of our Country, embracing Chapters on Bees and their Parasites, Moths, Flies, Mosquitos, Beetles, &c.; while a Calenda gives a general account of the more common Injurious and Beneficial Insects, and their Time of Appearance, Habits, &c. 200 pp Profusely Illustrated. 12 mo., cloth. Price $1.50.

WYMAN'S Ancient Inhabitants of Florida
Fresh water Shell Mounds of St. John's River, Florida By JEFFRIES WYMAN. Royal Svo. pp 94. With 9 plate Paper. $2.00.*

This is one of the most important works ever published on American Archæology. It relates to the implements of stone and bone pottery and habits of the ancient race of Florida, and will interes all collectors of Indian relics, and students of American Archæology.

www.ingramcontent.com/pod-product-compliance
Lightning Source LLC
Chambersburg PA
CBHW021821190326
41518CB00007B/689